RAISING HEALTHY HORSES

by

DR. BOB BEEDE

CONTRIBUTOR-DR. JACK AMEN

ILLUSTRATORS-DR. KELLY WARD JESSICA EVANS AND SHANNON SMITH

Christian Veterinary Mission

A Publication of Christian Veterinary Mission

www.cvm.org

Permission requests should be addressed to:
vetbooks@cvm.org

OR

Christian Veterinary Mission
www.cvm.org

FIRST EDITION—Revised

This is a first edition of Raising Healthy Horses. We think it will meet a need in many countries where horses are still used for herding, tilling, transportation and pleasure. In addition, we think it will have application in developed countries for those with pleasure horses. The contents are accurate but are given in an easy to understand style with many illustrations to enhance its clarity. We hope that whatever your interest in horses may be, that it will meet your expectations. If you have suggestions or questions concerning contents of this book, please address them to CVM, Seatlle using the address given on the previous page.

Raising Healthy Animals Series

Every year, thousands of people around the world struggle to survive because they don't have the right knowledge, skills and resources to care for their animals. Christian Veterinary Mission (CVM) sends veterinary professionals to live and work alongside many of these people to encourage them and provide them with not only much needed veterinary expertise, but also the hope that is only found in Christ. CVM veterinarians build lasting relationships with individuals and communities, helping them be transformed through Christ's love.

CVM, in its effort to be meaningfully involved in work in the developing world, quickly found there was little appropriate educational material available. CVM set about developing basic resource materials in animal husbandry for farmers and agricultural workers. Apparently, they met a real need, as these books have been accepted throughout the developing nations of the world.

The series of books published by Christian Veterinary Mission includes the following in order of publication:

Raising Healthy Pigs *	Drugs and Their Usage
Raising Healthy Rabbits *	Where There Is No Animal Doctor
Raising Healthy Fish	Raising Healthy Horses
Raising Healthy Cattle	Zoonoses: Animal Diseases That Affect Humans
Raising Healthy Poultry *+	Raising Healthy Honey Bees
Raising Healthy Goats *	Slaughter and Preservation of Meat
Raising Healthy Sheep	Disease and Parasite Prevention in Farm Animals

[Also available in: * Spanish + French].

CVM fieldworkers have also developed specific training materials for the countries in which they work.

All of these books have been put together by Christian men and women; in a labor of love and service, for people in need throughout the world. It demonstrates dedication to their profession, service to humanity and a witness to their faith. We hope that they are a help to you in developing an appropriate livestock program to meet your needs. We pray God's blessing on their use.

Leroy Dorminy
CVM Founder

TABLE OF CONTENTS

Section 1

Introduction

This book was prepared by Christian Veterinary Mission as an aid to those working in areas where information and assistance needed by small horse owners is limited and veterinary assistance is often not available.

As of the first printing, the editor of this book had been in private veterinary practice for 28 years in rural Idaho. The author has worked to educate the horse owner about practical methods of health care via numerous articles and lectures. He has also made several short-term missions for Christian Veterinary Mission to Peru.

This book is very brief, written in general terms, and is intended only as a reference manual. The book is not intended to replace the services of a veterinarian if one is available. Further detailed information can be obtained from the listed reference materials and from agriculture specialists or veterinarians within your residing country. Specific local conditions may not match suggestions listed in this book in allcases. You should therefore consult local authorities. **Any suggestions concerning improvements or future inclusions are welcome and appreciated.** Send or e-mail your comments to Christian Veterinary Mission or the author. Thank you.

Section 2

Normal Horse Basics

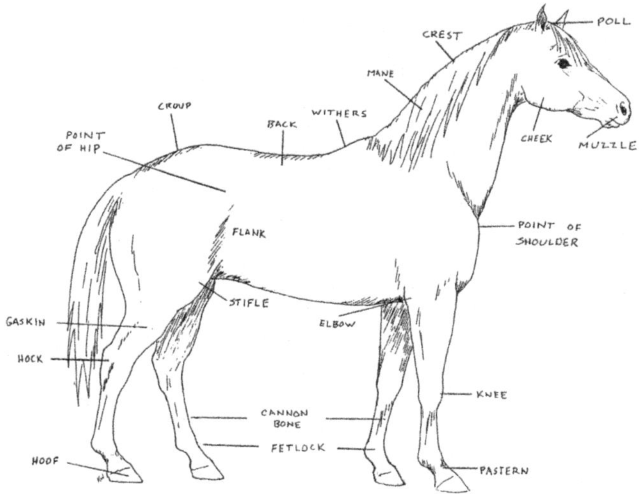

PHYSIOLOGICAL DATA
(Normal Body Functions)

Rectal temperature:	99 - 101.5 degrees F. 37.5 - 38.5 degrees C.
Heart rate:	40-48 beats per minute average.
Respiration rate:	10-14 respirations per minute.
Intestinal motility:	Mixing sounds from the gut are 2-5 seconds in duration and 2-5 per minute. Propulsive-retropulsive gut sounds are15-30 seconds long and occur every 2-5 minutes. These are gurgling type sounds heard over the flank both high and low on each side of the body.
Puberty:	Fillies may cycle at 1 to 2 years of age. If their development is slow or delayed due to poor conditions, it is best not to breed them until they are 3-4 years of age.
Heat cycles:	In estrous for 5-7 days, out of estrous 14 days. Total cycles lasts 21 days. A mare cycles during days of increasing sunlight and is anestrous (no heat cycle) during days of decreasing sunlight length.
Pregnancy:	Length is 325 to 365 days with 345 day average.
Age:	The central incisor tooth erupts at 2 1/2 years, the middle incision at 3 1/2 years, and the lateral incisor erupts at 4 1/2 years of age.

RECOMMENDED MEDICAL SUPPLIES AND EQUIPMENT

_____ Rectal thermometer (keep cool, out of sun)

_____ Stethoscope

_____ Twitch

_____ Halter, lead rope

_____ Flashlight

_____ Dental floats or rasps

_____ Wound cleaning kit
- surgical soap
- betadine disinfectant
- gauze sponges
- forceps

_____ Wound treatment
- betadine ointment
- fly spray
- Epsom salts
- scalpel & blade
- insecticide mixed with Vasoline

_____ Bandage material
- telfa pads
- cotton padding
- vetwrap or elasticon
- adhesive tape

_____ Eye care
- saline eye wash (buy, or mix 1 teaspoon of salt in 1 pint of boiled water to make)
- ophthalmic antibiotic ointment (nonsteroidal)

_____ Hoof care

hoof pick or knife
hoof nippers
hoof rasp
small hammer
shoes and nails if needed
bleach or lysol
Epsom salts
iodine

_____ Medication

Penicillin
Needles and Hypodermic syringes
Phenylbutazone
Dipyrone or Banamine (Flunixin Meglumine) for colic

_____ Parasite products

Ivermectin (Eqvalan or Ivomec)
Pyrantel Pamoate (Strongid)
Oxibendazole (Anthelcide)
Febendazole (Panacur)

_____ Vaccines

Tetanus Toxoid
4-Way (encephalitis, tetanus, flu)
Rhinopneumonitis (Herpes Virus)

EXAMINATION CHECKLIST

A. History - A thorough history of appetite, feed, schedule of feeding, activity, attitude, lameness, disease problems, manure, urine, etc. should be obtained. Determine if animals have been transported recently and if they are from a different area.

B. Clinical Exam - This exam should establish the current state of health and include the patient's attitude and evaluate each body system including the nervous, circulatory, respiratory, locomotor (bone, muscle), digestive, urinary, reproductive, and integumentary (skin, hoof) systems.

1. **Observe from a Distance** - Note general condition, stance, behavior, respiration, appetite, movement, any swellings, any discharge and any skin problems.
2. **Observe the Surroundings** - Note the pasture, water availability, roughage fed, condition of the other animals, signs of struggling in the dirt, etc.
3. **Lameness & Movement** - Note any lameness, especially when turning in a circle and trotting. Movement on a hard surface will help show some lameness.
4. **Appetite** - Loss of appetite is one of the earliest signs of illness including colic, infection, or disease.
5. **Rectal Temperature** - Shake the thermometer below 94 degrees Fahrenheit or 34 degrees Celsius and insert into the rectum. Check the temperature after three minutes. Fever is a higher than normal body temperature and indicates the body is fighting an infection. However with exercise, excitement, or hot, humid weather, the body temperature may increase without an infection.
6. **Heart Rate** - Listen behind the left elbow for the lub-dub of one heart beat. Count the beats over one minute. Rates over 60 can indicate pain as in colic episodes.
7. **Respiration** - Watch the side, count the respiration per minute. Note if the breathing is labored or if coughing occurs. Watch the nostrils for any discharge or flaring (as if fighting for breath). A respiratory rate above 16 can indicate a respiratory problem or anemia. Coughing and nasal discharge may indicate allergies, infection, or choke.
8. **Membranes** - Examine the color of the gum (should be pink). Press on the gum and then remove your finger and count the seconds before the gum turns pink again. The capillary refill time should be 2 seconds. Pale membranes and/or slow capillary refill time can indicate anemia, shock, colic or blood loss. Dark red membranes indicate endotoxic shock. Blue or gray membranes indicate

imminent death. Yellow membranes indicate blood cell destruction or liver problems.

9. **Gut Sounds** - If the horse is showing signs of colic, listen with a stethoscope on each flank, high and low, for GI motility. Gas will sound like wind chimes. Lack of sound indicates the intestines are not moving.
10. **Mouth and Teeth** - Drooling and bad mouth odor can indicate a foreign body is stuck in the mouth or there is an infection of the tissues. Examine all sides of thetongue for cuts and foreign bodies. The cheek teethshould be smooth on the surface. Sharp points on the lower inside and upper outside of the cheek teeth indicatethey need to be floated or rasped smooth so the patient can chew better.
11. **Eyes** - Drainage from the eyes can indicate fly problems, infection, obstructed tear duct, or foreign bodies such as grass awns. If the patient squints, there may be an ulcer on the cornea (outside covering of the eye). Also examine the membranes for color. Determine if the animal can see or is walking into things. In some warm, moist climates, a parasite can invade the surface of the eye and the horse will need Ivermectin treatment orally.
12. **Feet & Legs** - If the horse is lame, begin by determining which leg or legs are involved. Look at each hoof, taking care to clean it out, looking for punctures, nails, etc. Feel each limb for swelling, heat, punctures, or discharge. Use a small hammer and gently tap the hoof looking for sensitive areas. Flex a sore limb for 2 minutes and then trot the patient. If the flexed area has a problem, the soreness will be exaggerated for several steps.
13. **Urogenital** - Examine the vagina or sheath. Look for any discharges and also note any diarrhea.

14. **Udder** - Examine the udder of all nursing mares or jennies to determine if they are producing milk of adequate quantity and if there are any swellings which might indicate an infection. The milk should be white, not bloody or chunky.
15. **Hydration** - Use the skin pinch test to determine the degree of dehydration if the animal is sick. Lift a pinch of skin on the neck then release. If the skin ridge persists for any length of time, the animal needs fluids.
16. **Skin & Coat** - The condition of the skin and coat can indicate if parasites could be present, if the diet is poor, or if a disease is present.

C. **Diagnostic Tests** - If some lab service is available, diagnostic tests may be helpful. Fecal exams under the microscope will determine parasite load. Complete blood tests will help determine anemia, infection, and disease. Blood chemistry can determine if organs are involved. Uterine (womb) cultures will determine the best antibiotic to use if the reproductive tract is infected. However, most primitiveareas will not have the services available or the cost will be prohibitive so educated guesstimates may suffice.

Section 3

Environment
Behavior
Restraint

ENVIRONMENT

Confinement - Horses, mules, and burros can be confined by fences, by tethering with a rope to a stake, or using front leg hobbles which can be made of burlap bags. They do need protection from the heat and cold. This protection can involve the use of natural windbreaks like trees and shrubs, rock walls, and buildings. Young animals especially, need protectionfrom cold winds, rain, and intense sun. However adequate exercise, access to pasture or roughage feeds, and water is very important if the animals are strictly confined. Strict confine- ment can lead to boredom and the development of vices suchas eating dirt, cribbing (chewing on fences), wind sucking,weaving, kicking, etc.

1. **Tethering or Picketing** - Attach a rope to a halter or around the neck in such fashion that the loop will not slip and choke the animal. A bowline knot (p. 24) works well to accomplish this procedure. The tether rope is best made out of soft fibers like cotton or wool so it will not cause burns (nylon is harmful) if wrapped around the limb. Attach the rope to a stake, a rock, or a tree. The animals should be moved to water at least twice daily and moved to good grazing areas daily. By rotating the tethered areas daily, picking up the manure (compost for fertilizer), and not going back to that area for 6 weeks, the parasite loads can be reduced. Care must be exercised to avoid ropeburns around the pastern or the animal becoming tangledup in obstacles perhaps breaking a leg or hanging itself.

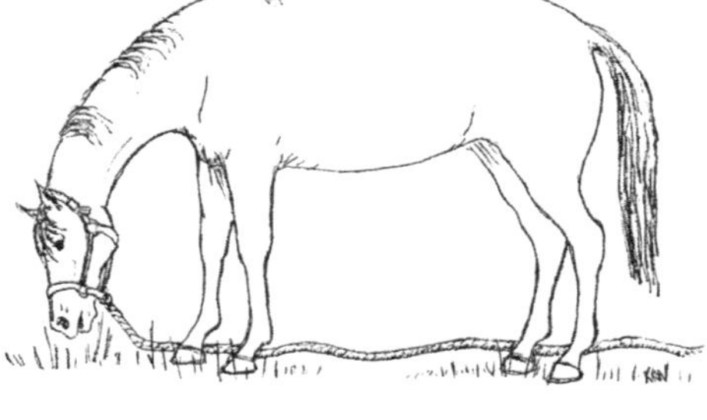

2. **Tethering or Picketing by the Leg** - This method is safer to use than picketing by the neck. There is less chance of injury if the animal becomes tangled up. Attach a hobble, a burlap bag hobble, or a large soft cotton rope to one front leg at the pastern. The hobble or rope around the leg must be loose and not able to be tightened to avoid tourniquet action on the limb if the horse pulls hard on the picket rope or becomes tangled. Then attach the picket rope to a stake.

3. **Open Grazing** - Open grazing will allow the animal to seek shade, shelter, water, and feed when needed. Problemscan result from over-grazing pastures, lack of pasture rotation, grazing in cultivated areas, or the animal escaping.

4. **Hobbles** - A burlap bag can be used to fashion hobbles that fit around the front pasterns and keep the horse from running off. This form of confinement is inexpensive when compared to fencing and the animal can be herded to desired pastures. Hobbles left on too long can result in stiff or sore pasterns or result in heel cuts. Side lines at- tached to the left front and the left rear leg can also be ef- fective ways of confinement. Again, care must be takento avoid constriction of the pastern by the hobbles or side- line ropes. See pictures on next page.

Burlap bag hobbles: Remove seam so that both ends of burlap bag are open. Then gather together and twist 3 times.

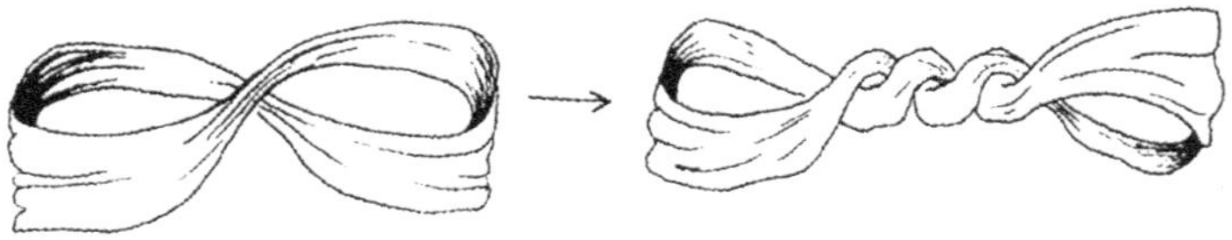

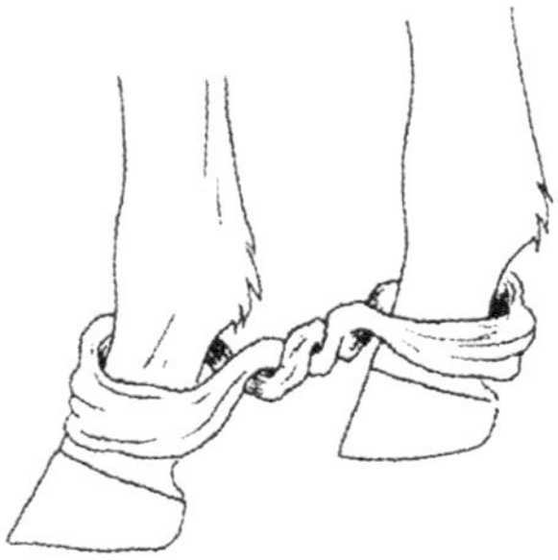

5. **Pens** - This is a good confinement method, but the animal must be given feed and water each day. Dry roughage in the form of hay may not be available or affordable. The pens should offer shade and shelter in addition to fresh water. Again, strict total confinement may result in vices so some form of movement or exercise is important.

Pasture Management - Pasture grazing should be rotated to avoid over grazing and buildup of parasites. It is desirable to rotate the pasture area weekly and avoid going back to the original pasture for 6 to 8 weeks if possible. All manure should be picked up and composted for fertilizer in cultivated tracts. This rotation will greatly reduce the parasite load on the animals.

Grass pasture is best for horses and mules. If the grass is very rich and lush the animal can become overweight and develop founder or laminitis (see hoof conditions—Section 9). If the animal is getting too fat, confinement to a dirt pen or area for part of the day and shorter grazing intervals will be helpful.

BEHAVIOR & RESTRAINT

Behavior

1. **Reflex Behavior** - The horse, burro, and mule exhibit a reflex behavior that enables them to escape from potentially harmful situations. This reflex is the flight or fight reaction. These harmful situations are recognized by the animal through its senses of sight, sound, taste, odor, and touch. By this reaction, each animal is able to maintain its security. For example, animals approached from their visual blind spot directly from behind may kick or suddenly jump if startled. Care and caution should be taken when working with any equine. Let them know you are approaching by the reassuring sound of your voice and approach them from an angle where they can see you.

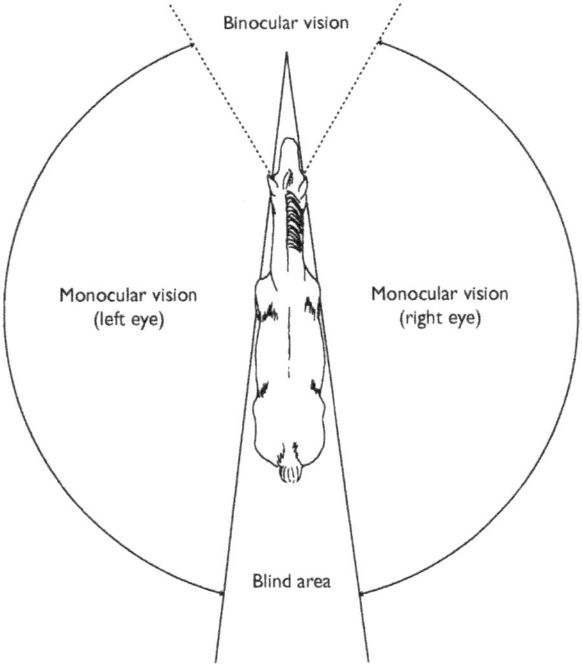

2. **Dominance Behavior -** Because horses, mules, and burros are herd animals they have developed a pecking order with social rules and a hierarchy. Dominance is based on the self-confidence of each individual as much as on body strength. Squabbles about rank rarely develop into a full-fledged fight. Usually an aggressive posture is sufficient although a kick, bite, or strike may be used to stress a point. Humans also enter into the pecking order as a tame horse regards man as one of its own kind. By being self-confident and if necessary, aggressive we may play the role of the highest ranking "horse." This is the clue to correct handling of the horse.

3. **Caution, Problem Horses -** Horses may strike out with their front feet when they are being treated or restrained. Also remember that these patients can kick to the side with their hind feet as well. Care should always be taken to standclose to the side of the horse, not in front or behind the animal nor to the side of the hind feet if they tend to kickout sideways. Some mares with foals and stallions can viciously bite or use their head to push you away. These animals may approach you with their ears pinned back.Always be alert and use common sense when working around equines. Animals can detect if you are fearful. Approach them with confidence or find someone who is not

afraid. It is also best to stand close to the animal instead of a half meter away. If the animal kicks when you are close, they will merely move you away. If you stand several feet away, you are more likely to be injured by a kick. Again, have an escape route when working with these animals but beself-confident as you deal with them.

4. **Submission of the Equine -** The horse, mule, or burro depend upon sprinting away from danger in order to survive. The defensive position is to stand with legs squared, head up, ready for flight. The position of greatest vulnerability for the wild horse is when it is feeding or drinking. He cannot see, hear, or smell as well with the head lowered. The equine therefore signals submissiveness by lowering the head. In addition, they may also signal vulnerability by making feeding motions (chewing or licking the lips). This action is most common with young foals. Flight deprivation is how we make the horse dependent on us and create an attitude of submission. See the section on restraint (following) for more information.

5. **Pecking Order Management -** Those individuals at the bottom of the pecking order may need extra feed or separate confinement to avoid injuries and to ensure adequate feed intake. Never feed dry forage to groups of animals where one can be trapped in a corner when trying to eat. If possible, always spread the feed out over a large open area to avoid injury. Add several extra piles compared to the number of animals to avoid any animal from being deprived of feed.

Training and Restraint - Restraint is the art of dominance. Some individuals will be very submissive and need little restraint. Other more dominant individuals will require varying measures of restraint. It is very important to understand that we cannot physically dominate the horse, mule, or burro. Dominancealso does not refer to physically abusing, hurting or mistreating them. We should never inflict pain or frighten these animals. Instead we need to create submissiveness by making the animal feel dependent upon us—respectful and obedient, not fearful.

Most techniques of horsemanship involve the principle of flight deprivation. Halter breaking the horse if done correctly conveys to the horse the feeling that flight is impossible. Using round pens to train or small pens to treat the equine works because they have learned escape is impossible. Animals broke to hobbles become more submissive. Flight deprivation creates dependency, respect, affection and a desire to follow that person's leadership.

Restraint Techniques - The method of restraint depends on many factors including the animal's temperament, age, size, physical condition, procedure to be performed, equipment,help and drugs available. It is important to use humane techniques and alleviate pain as much as possible plus prevent further injury to the animal.

Individual horses, mule, and burros may react differently in each situation. The best method of restraint may be a matter of experimentation. Apply the least restraint initially and then increase in the order of severity. If available and the animal's condition permits, sedatives like acepromazine or xylazine are preferable.

1. **Temporary Rope Halter -** Use a long rope. Place a loop around the animal's neck using a bowline knot (p.24) Pass a bight in the standing part through the loop and over the nose. Pull the nose piece tight when the halter is used. Never use a rope around the neck that will tighten and possibly suffocate the animal or damage the nose!Caution! Many animals will rear back or strike out with a front foot if pressure is too great.

(Courtesy of Restraint of Animals.)

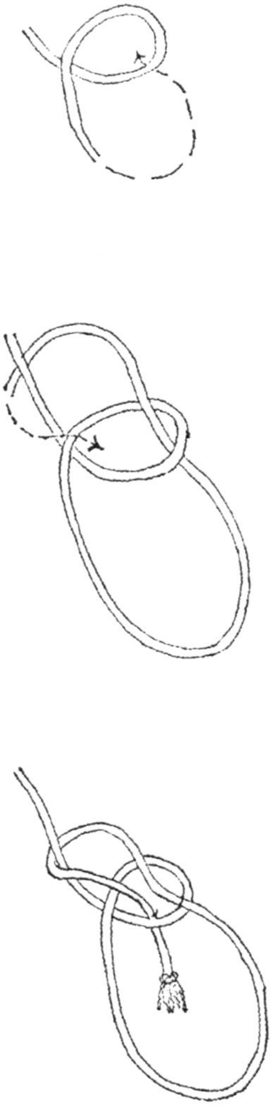

Tying a bowline. It will not slip but will almost always untie easily when not under a pull. Used for tying a rope around a horse's neck or foot. It will not slip tight and injure or choke the horse.
(Courtesy of Western Horse Behavior and Training.)

2. **Halter and Lead** - The most important assistant is the person controlling the animal's head. The assistant should stand on the same side of the horse as the operator to revent kicking. Small chain shanks are helpful with spirited animals. The chain can be applied over the muzzle, under the jaw, through the mouth, or over thegums under the upper lip. Take care not to be abusive with the use of the chain and do not tie them when a chain is used. It is also not advisable to tie a horse by the lead rope and attempt to examine it. These individuals may sit back on the rope and break the equipment or lunge forward and injure themselves or the handler...the idea is not to hold the animal but to control its actions.

3. **Skin Twitch** - The skin twitch is applied by grasping a roll of skin over the shoulders with one or two hands. This may work well with some horses that strike.

4. Ear Hold - Grasp the left ear firmly with the right hand. The ear is squeezed and slight pressure is applied. The right forearm should be firmly against the animal's jaw to avoid injury to the holder if the horse rears or swings its head. Do not twist the ear or use a twitch on the ear as you can damage the cartilage.

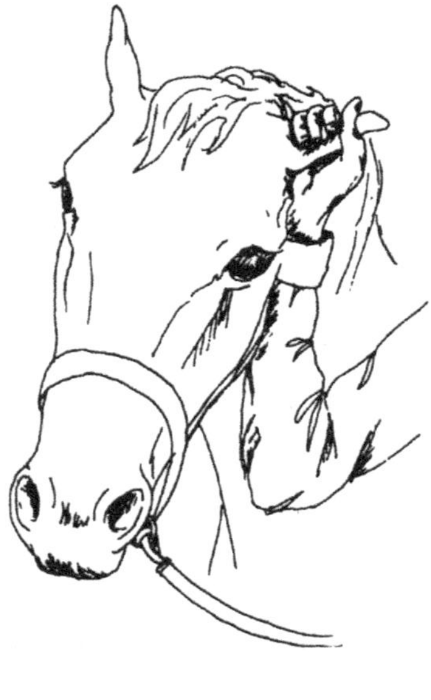

5. **Hand Lip Twitch -** Grasp the upper lip with your thumb and forefinger and pinch the nose. Take care not to stand in front of the animal as it may strike with the front feet.

6. **Rope or Chain Lip Twitch -** Use an long stick, 2 feet in length and at least 2 inches in diameter. Put a hole through the stick near the end and attach a large string or small cord through the hole. Tie the string ends together so a loop of 3 or 4 inches remains. This twitch can then be tightened on a pinch of the horse's nose by twisting the stick or handle. Care must be taken to not tighten thetwitch too tight. Some animals will strike at the twitch so be sure to stand next to the shoulder when using this form of restraint.

7. **War Bridle** – A rope or cord with an eye or loop on one end is needed. The other end is passed through the eye to form a small loop. This is placed around the lower jaw so that the rope passes across the inter-dental space and beneath the tongue. The end of the cord is carried around the top of the head (the poll) and down the other side of the face, crossing beneath the loop around the jaw. The end may be pulled down when restraint is desired.

(Courtesy of Restraint of Animals.)

8. **Blinds** – This blind is useful in loading a shy horse on a truck or controlling him or her in other situations in which they might become unmanageable.

(Courtesy of Restraint of Animals.)

If more protection seems needed, the padded hood or a thick folded blanket may be used.

(Courtesy of Restraint of Animals.)

9. **Leg Lift -** Lift and flex a leg to keep an animal from jumping while working on it or use a rope to hold the leg up.

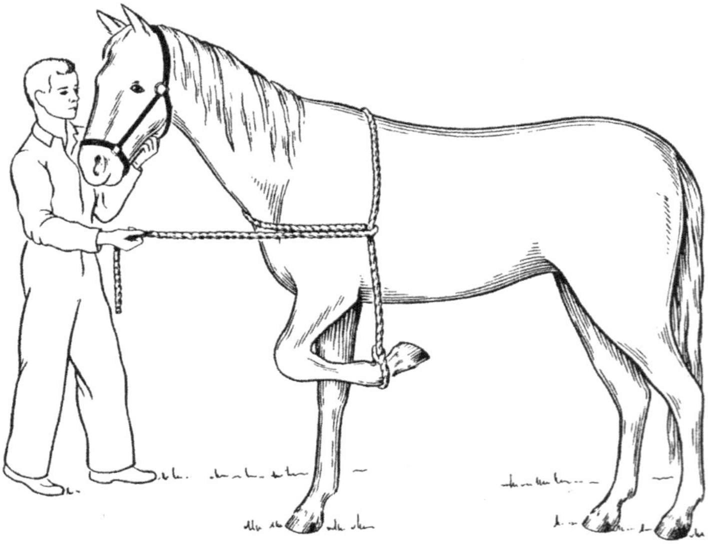

(Courtesy of Restraint of Animals.)

10. **Front Leg Hobble -** Tie the front leg up in a flexed position with a burlap bag. After the horse has struggled with this restraint for 10 minutes or so, they will often be very submissive. This is an excellent method to make strong-willed animals submissive.

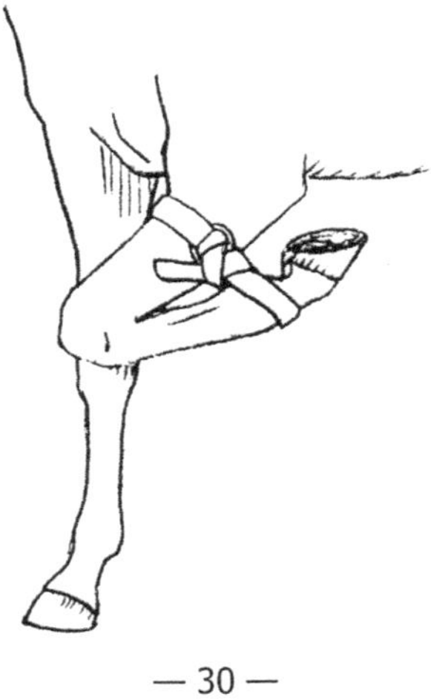

11. **Body Cradle &/or Tail Restraint -** This method works well on burros and foals. Stand to the side of the body to avoid being kicked. Cradle the body at the shoulders and rump. If there still is difficulty restraining the foal, gently grasp the tail at the base and pull straight over the back. This technique tends to steady and calm some animals.

(Courtesy of Restraint of Animals)

A rope looped behind the hind quarters is a good way to lead a foal or reluctant burro.

12. **Tie Hind Leg Up** - Make a loop around the neck using a bowline knot. Take the standing rope and pass a bight through the neck loop and down to the pastern on the hind leg. Slowly pull the hind leg forward with the rope coming through the neck loop. Tie this leg up with a slip knot. This restraint will help prevent the animal fromkicking with the rear legs.

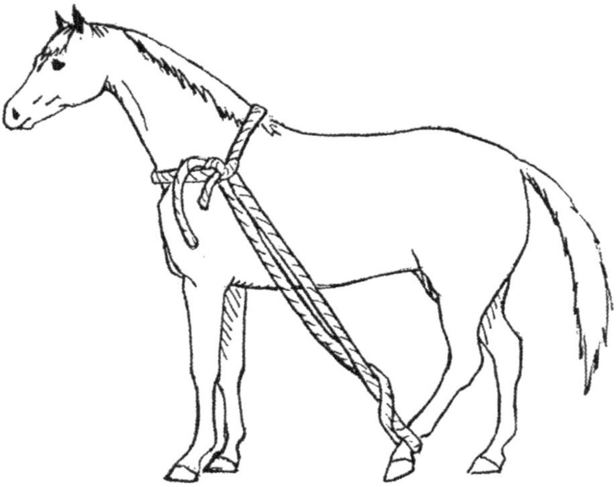

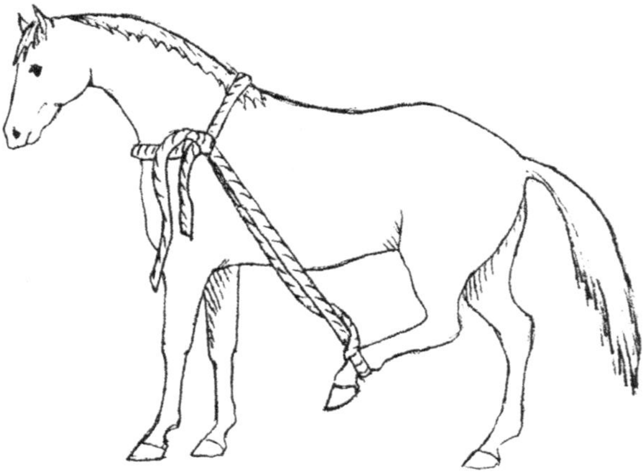

13. **Breeding Hobbles** – Breeding hobbles are used to prevent a mare from kicking a stallion. These hobbles are also useful to restrain both rear legs for rectal or vaginal examination. Follow the diagrams on pages 85 and 86 to make such a hobble.

14. Casting a Horse - If you need to throw an animal to the ground this approach may be helpful. Warn the owner that excessive struggles can lead to injured limbs so this procedure must be used only in extreme circumstances.

Double Side-line Casting Harness

The double side-line casting harness is the most practical method of casting a horse and restraining him in the recumbent position. It gives a maximum of restraint without special equipment, only two pieces of thirty foot rope, or one sixty foot piece being required. If web hopples are used, as illustrated, they will lessen the chance of rope burns on the animal's pasterns.

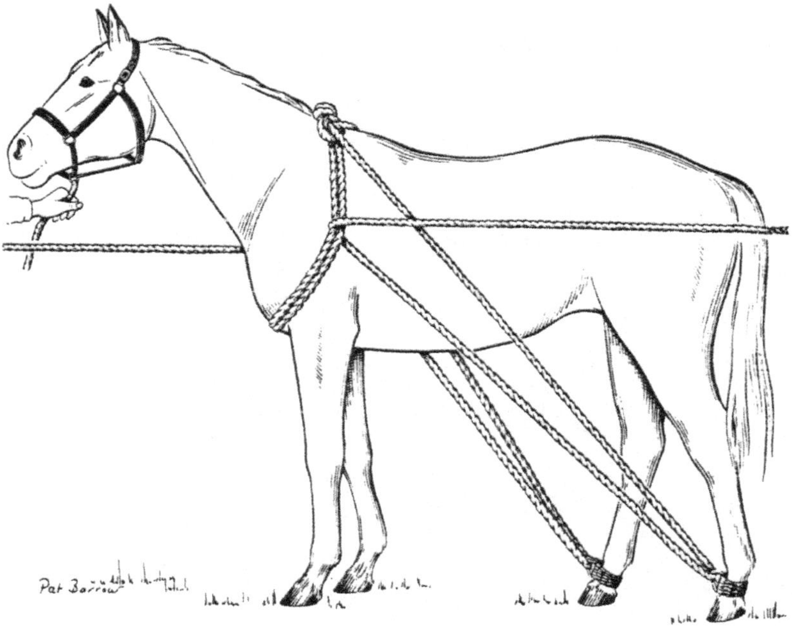

Double a sixty foot rope or tie together two thirty foot pieces. Form a rope collar that fits around the horse's chest, shoulders, and withers. A bow-line on a bight is a convenient knot to use in forming a single strand rope collar which can be passed over the horse's head and fitted in position. A square or bowline knot can be used to form a double strand collar in position around the

(Courtesy of Restraint of Animals.)

horse's shoulders. The ends of the rope should leave the collar at the withers, pass down, one on each side of the body, and go through both rings of a web hopple that has been placed on the pastern of the rear leg of that side. Take the end of the rope on each side forward again to the shoulder and pass it, from back to front, beneath the collar.

The end of the rope on the side on which the horse is to be cast is carried forward and held by one or two men. The end of the rope on the opposite side is carried backward, around the rump, and away so that it forms a right angle with the long axis of the horse's body. One man holds this rope. The ropes must be held free of slack but without tension, or the horse will tread around in an effort to free his legs of the pastern hopples. One man holds the halter rope, which may be passed over the horse's head, at the poll, for leverage when pulled toward the side on which the horse is to fall.

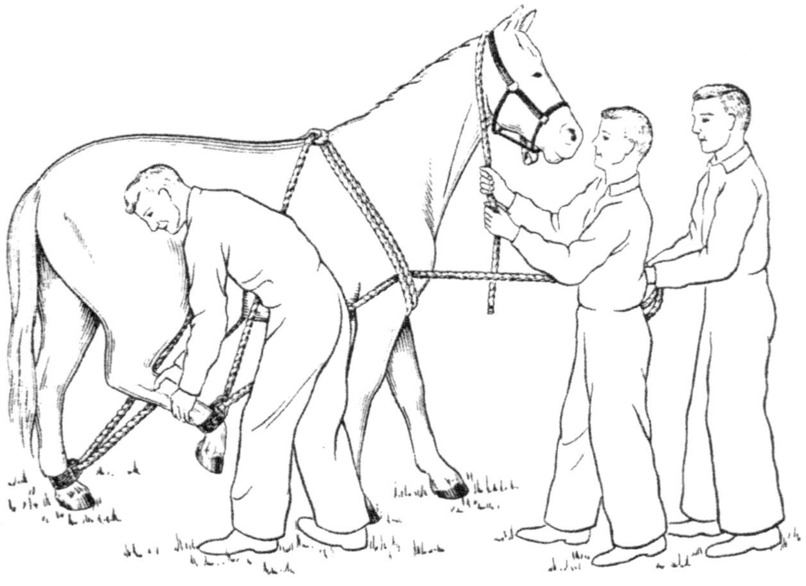

(Courtesy of Restraint of Animals.)

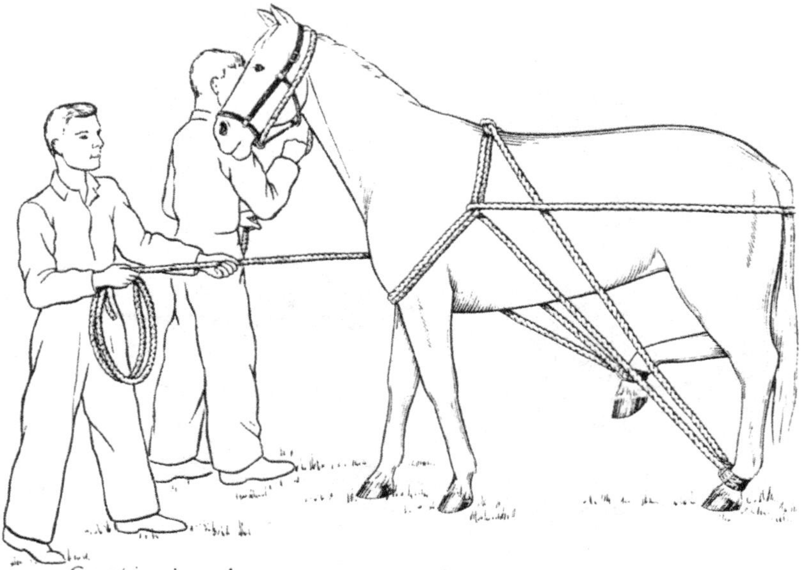

Casting: As one man raises the rear leg of the side on which the horse is to fall, the rope on that side is drawn tight to hold the leg in the raised position. The man who raises the leg moves clear of the horse, all the ropes are pulled simultaneously, and the horse is thrown to the ground. The man holding the halter rope prevents the horse from raising his head and arranges the bedding beneath it. Both the casting ropes are held tight, and the man holding the upper one moves up to the horse's croup in readiness to tie the rear leg.

Alternate Method: The halter rope is not passed over the poll as illustrated. Instead, the man holding it grasps it close to its attachment and with it forces the horse backward into the harness. As the horse lifts his legs in moving back, tension is maintained on the casting ropes. The horse automatically falls in the direction of the stress on the rope around his rump. In this method of casting the position of the men is as illustrated.

(Courtesy of Restraint of Animals.)

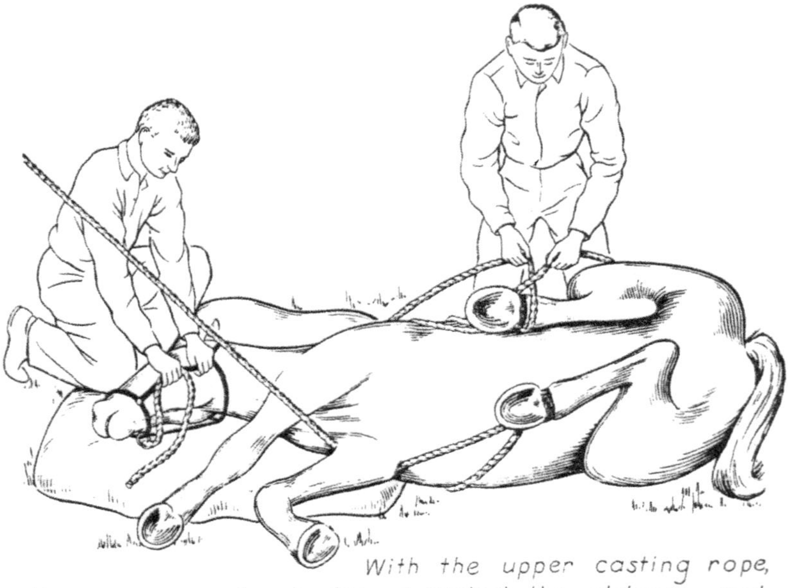

With the upper casting rope, the upper rear leg is flexed against the abdomen and a loop is thrown around the pastern from front to back.

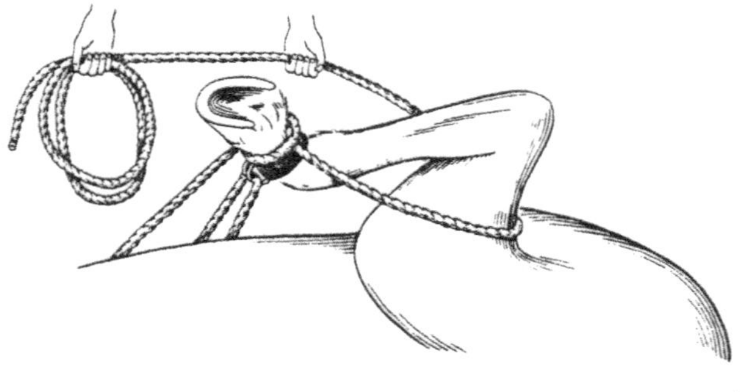

Tying is continued by passing the rope backward diagonally across the cannon bone, then across the medial side of the hock, and around the hock to the lateral side.

(Courtesy of Restraint of Animals.)

Pass the rope forward from the hock across the back of the cannon bone to the medial side of the pastern; then around the front of the pastern to the lateral side. Continue with the tying of the leg until two "figure 8-s" have been completed and the rope is at the lateral side of the pastern. Form a bight in the rope.

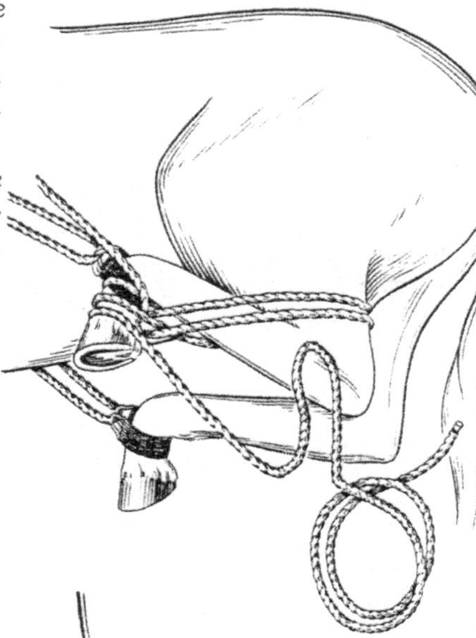

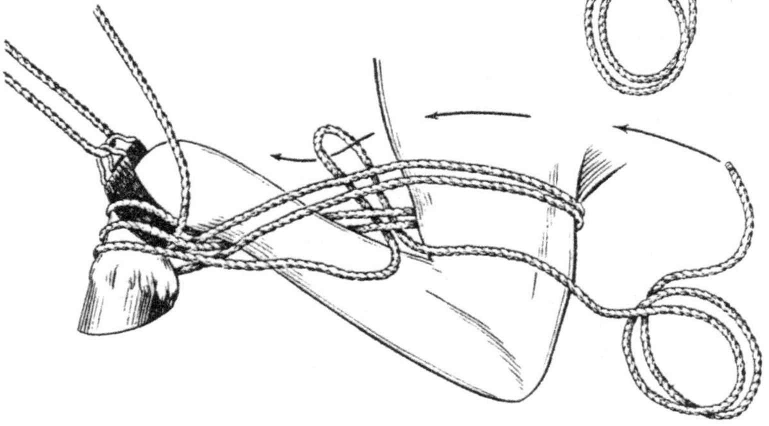

Pass the bight down between the ropes around the hock and the hock joint. Draw all the free end of the rope through the eye of the bight. Pull tight.

(Courtesy of Restraint of Animals.)

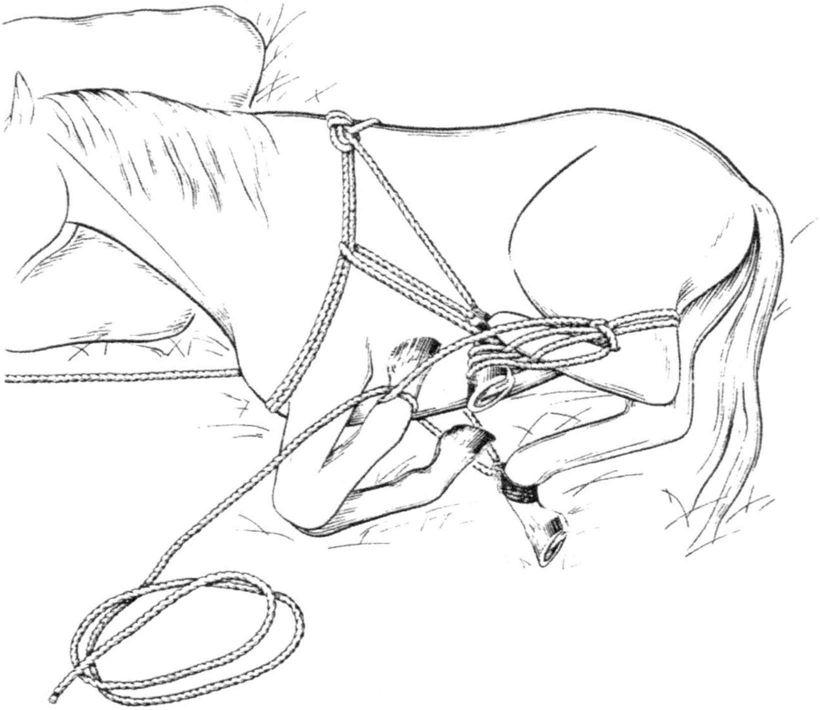

Still pulling tightly on the end of the rope, straddle the neck of the horse and face his body. With the left hand flex the front leg, and with the right hand throw a half hitch around the pastern. Pull the rope so that it draws the hoof of the front leg against the body and close to the hoof of the rear leg.

(Courtesy of Restraint of Animals.)

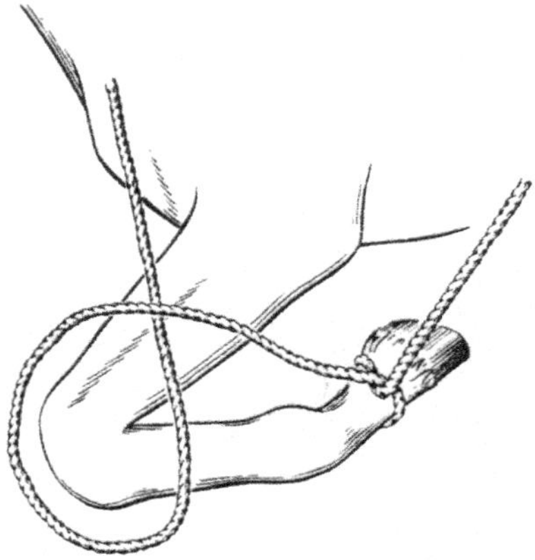

Form a half hitch in the rope and throw it over the flexed front leg.

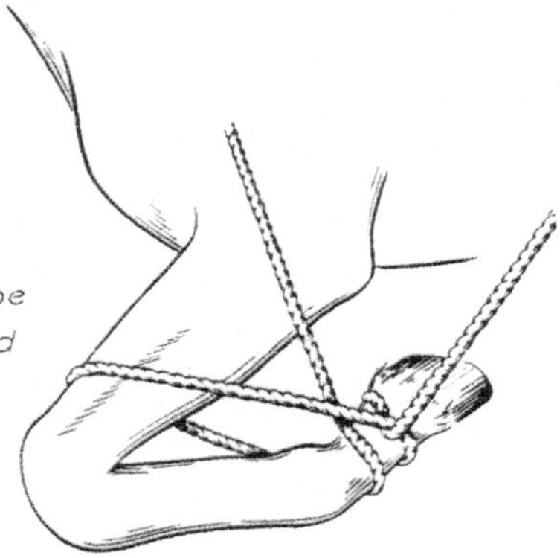

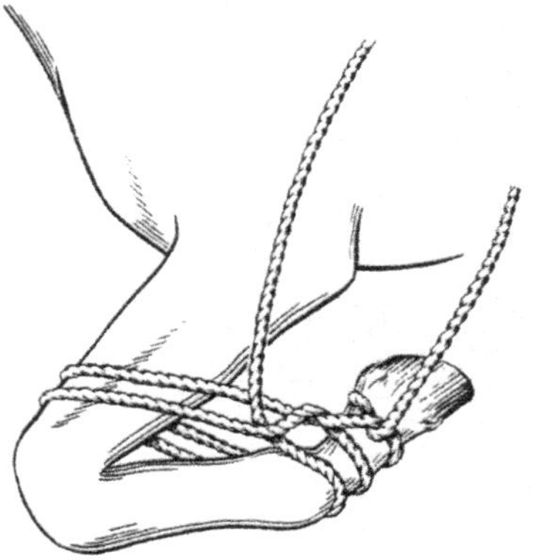

Throw another half hitch around the flexed front leg and draw the rope tight.

These illustrations, for the sake of clarity, do not show the leg in complete flexion as it should be when tied.

(Courtesy of Restraint of Animals.)

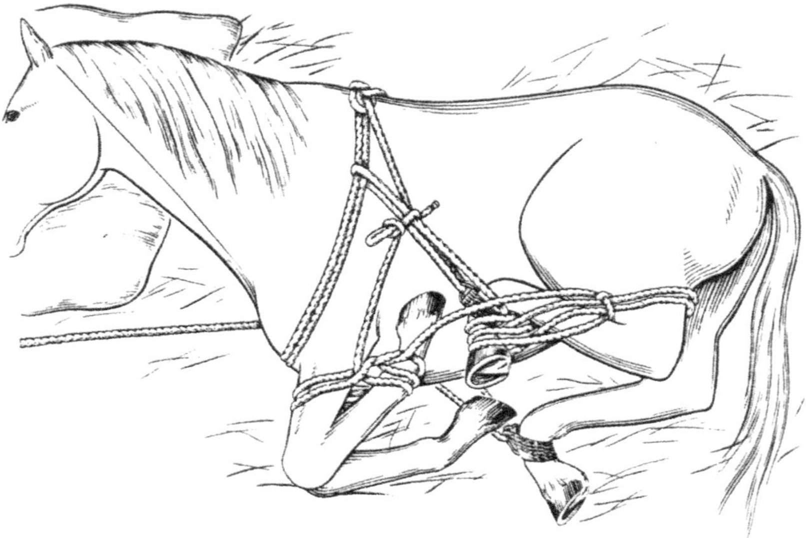

Carry the end of the rope over to the ropes that pass from the collar to the rear leg and tie it securely.

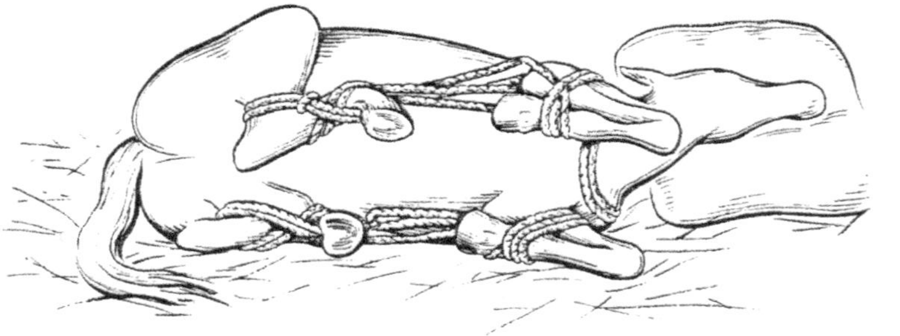

Continue to hold tight the rope which is beneath the horse. The man holding the head should shift his position, but continue to hold the head while the horse is rolled over and tied on the opposite side in a similar manner.

(Courtesy of Restraint of Animals.)

Section 4

Selection of Horses, Mules, and Burros

SELECTION

Selection of the equine depends on the future use planned for that animal. Riding and packing animals will need a different build than that of a draft animal used for plowing. Body form will dictate body function. Genetic defects are important to avoid in case you plan to breed any of the animals. Disposition and ease of handling are also critical to successful use of the desired animal.

Genetic Defects - These are traits that can be transferred from one animal to another with breeding. Depending on the genetic mode, not all offspring will show thedefect but they can carry the defective gene to an offspring of their own. Listed below are defects that should be avoided:

- Parrot Mouth - The front top and bottom incisor teeth do not match when the animal bites. This will lead to a situation where the animal cannot eat grass properly causing poor body condition.

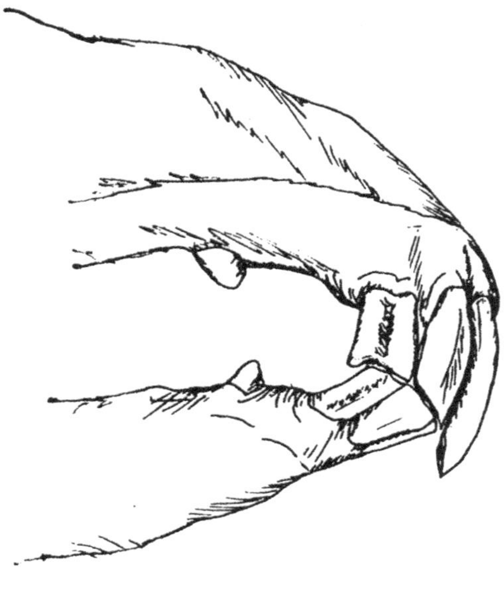

- Cleft Palate - the top of the mouth is not closed. This allows food to get into the respiratory area.
- Dwarfism - miniature animals that result from inbreeding. Many of these animals will have multiple problems or defects.
- Curved Spine - this is a deformity (misshaped, curved) in the vertebrae of the spine. The spine will not be straight.
- Crooked Limbs/Deformed Joints - this problem canbe genetic or due to injury or poor nutrition. If a work animal doesn't have good conformation they will not be able to do sufficient work.
- Cryptorchidism - one or both testicles are not descended.
- Hermaphrodite - mares with a vagina but also have a penis where the clitoris would be located. These animals are sterile and can be used for work, but the same breeding (father-mother) should not be repeated.
- Heart Murmurs - instead of a "lub-dub" heart sound,there is a hissing sound caused by a heart defect. These animals may not be able to perform hard work.

Conformation

1. **Eyes -** The eyes should be functional. Avoid blind or partially blind animals as they easily spook at objects and can cause injury to the handler.
2. **Feet -** The saying, "no foot, no horse" is very true.

- The front feet are a more rounded shape than the hind feet. The slope should be at an angle of 45 to 50 degreeswith the angle of each foot being the same. Avoid steep angled feet.
- The heels should be wide, the frog large and full toabsorb concussion. Avoid narrow donkey-like feet in horses, which do not provide enough bearing surface for the horse during hard work.

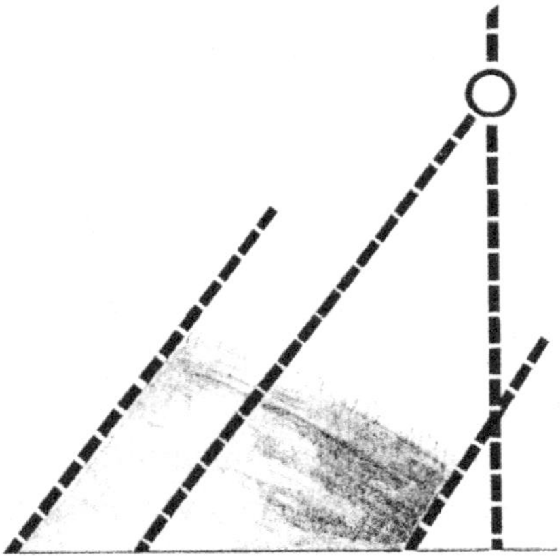

(Courtesy of Feet & Shoes.)

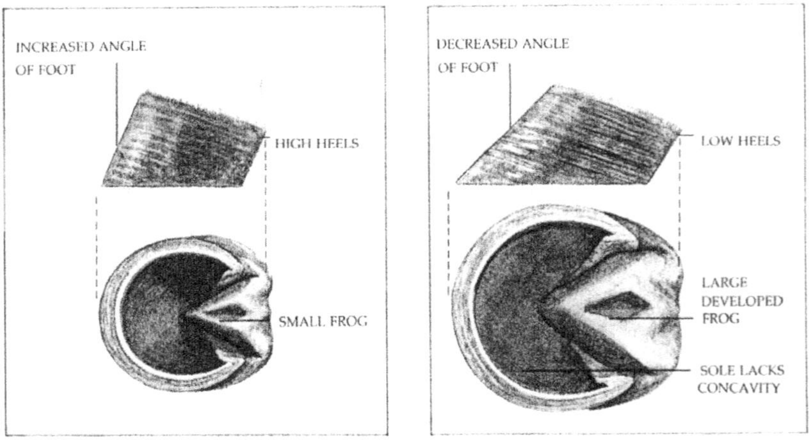

PEDAL BONE or **THIRD PHALANX**
– the principal bone of the foot

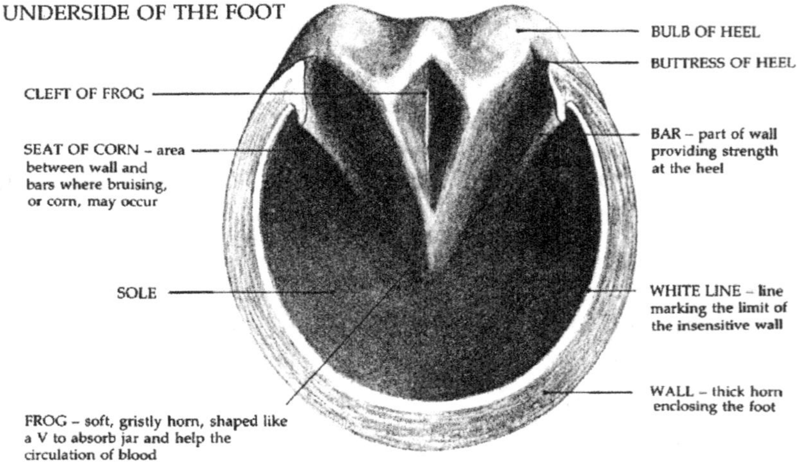

(Courtesy of Feet & Shoes.)

- The walls should be strong and not easily chipped or cracked.
- Avoid flat feet - a concave sole is preferred. These animals are more prone to bruising.
- The feet should not turn outward or severely inward. This conformation can result in foot and leg problems caused by an un-natural foot path (see diagrams).

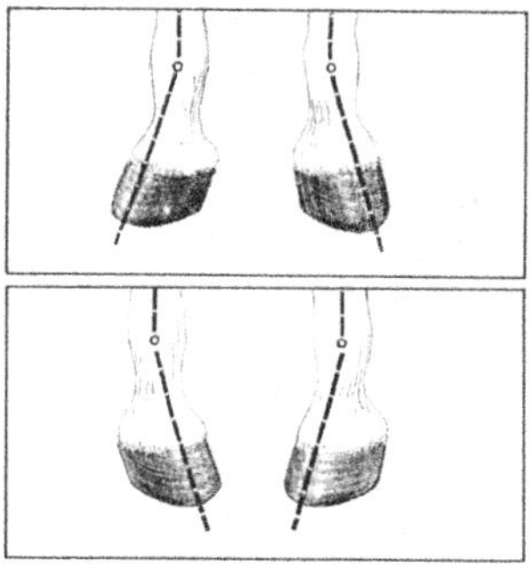

Toes which turn **out** (*top*) or **in** (pigeon-toes) are often allied to narrow or wide chests. Both kinds of feet cause uneven wear to the shoes.

(Courtesy of Feet & Shoes.)

3. Neck and Shoulders

- The neck and shoulders are very important in draft animals. Abnormal shoulder form may make fitting the harness difficult.
- Steep shoulders will make for jarring action when packing and riding.

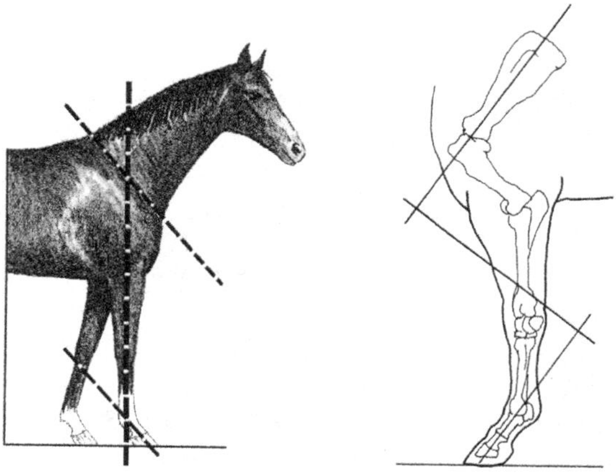

(Courtesy of Feet & Shoes.)

4. **Withers -** they should be well defined and of reasonable height. Those too high are prone to injury and those too low do not hold saddles well.

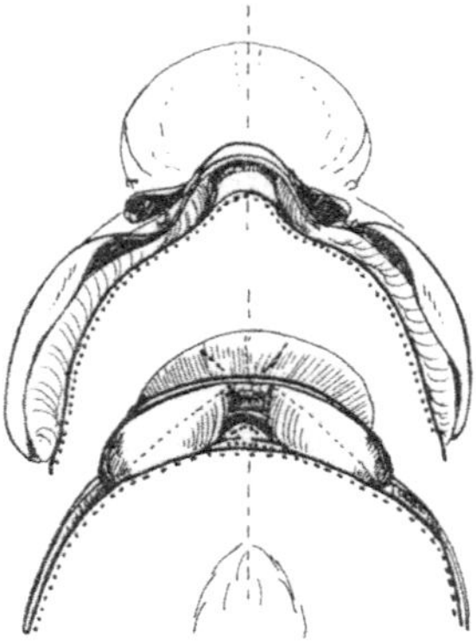

A saddle which fits correctly, seen from front and back.

(Courtesy of The Manual of Horsmanship.)

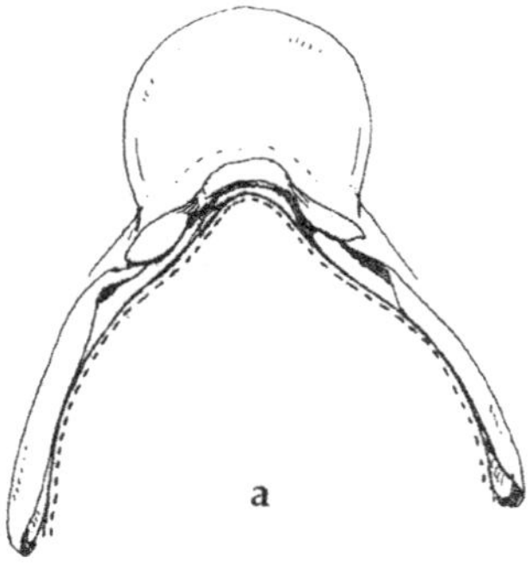

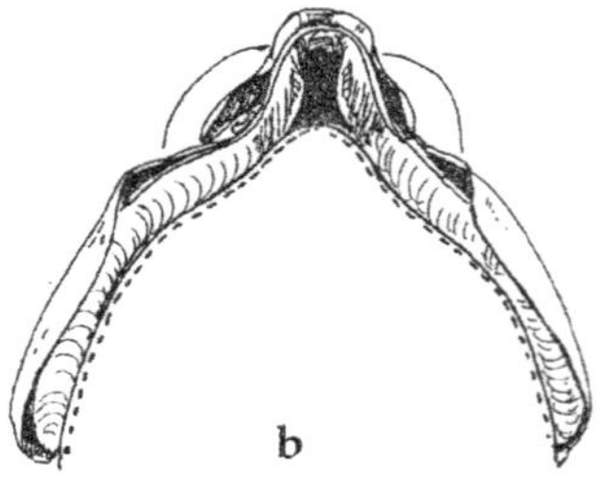

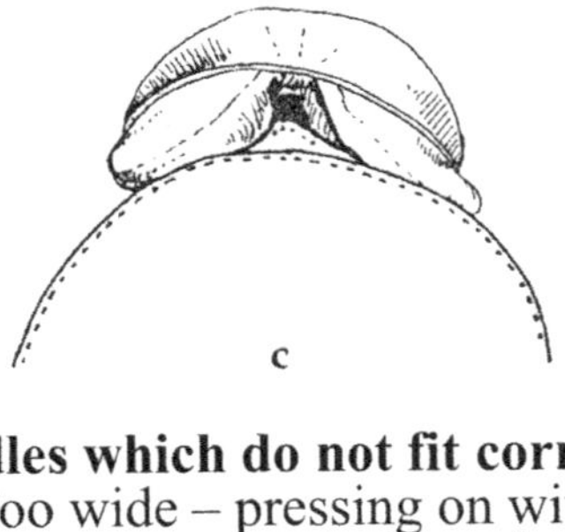

(Courtesy of The Manual of Horsmanship.)

Saddles which do not fit correctly.
(a) Too wide – pressing on withers and spine. Also, crooked.
(b) Too narrow – pinching the withers and spine.
(c) Sloping down on one side. (In need of stuffing).

5. Forelimbs

- Seen from the front the forelegs should appear to drop straight from the forearm to the foot with good bone below the knee.

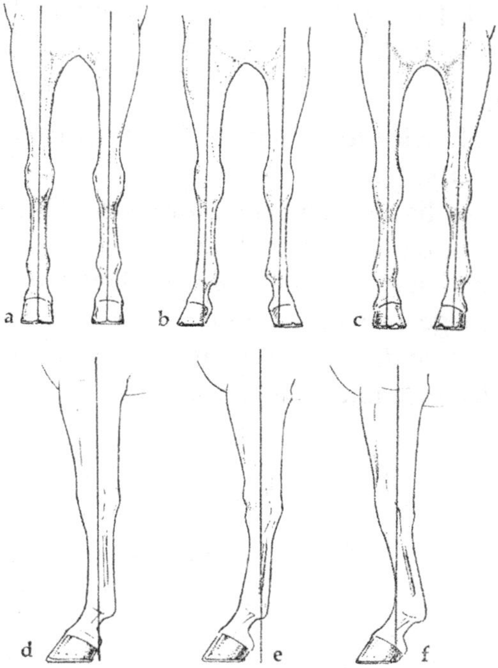

Forelegs. From front: **(a)** Correct. **(b)** Turned-out. **(c)** Turned-in. From side: **(d)** Correct. **(e)** Back at the knee (and open knee). **(f)** Over at the knee. *(Courtesy of The Manual of Horsmanship.)*

- The knees should be broad, flat, and deep from front to back. Avoid knees with indentations at the front, those presenting a concave appearance from the side, or anylateral deviation of the knee. Knees without symmetryor that do not flex well are a problem.

- The cannon bone should have a good circumference (good bone) for strength and be straight to avoid tendon strain.
- Ligaments of the limb should not show any bowingor swellings.

- The fetlock and pasterns should be medium length and a normal slope. Long pasterns are prone to strain. Upright pasterns absorb less concussion and can also make the animal prone to injury. Avoid any animal with reduced flexion, painful flexion, or swellings around the pastern (see ringbone).

6. Back

- Animals with long, narrow backs will not carry weight efficiently.
- "Hollow backs" have a dip or sway. This may be a sign of age or damaged vertebrae.
- "Roach backs" are humped up. This may also be a sign of damage or malformation
- Horses that dip or drop their backs when a load is applied may have an injury to the back

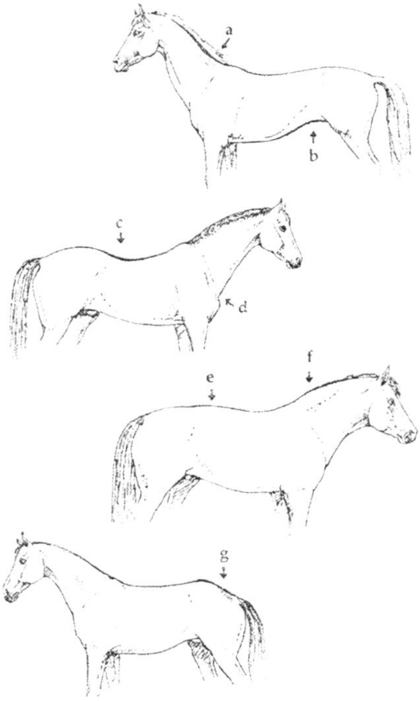

Conformation faults. **(a)** Ewe neck. **(b)** Herring gut. **(c)** Hollow back. **(d)** Straight shoulder. **(e)** Roach back. **(f)** Bull neck. **(g)** Goose rump.
(Courtesy of The Manual of Horsemanship.)

7. **Hindquarters**

- They should be strong and muscular as they provide most of the power in work.
- When seen from behind, the quarters should look rounded at the hip, showing a gradual swell of muscleon either side.
- Points of the hip should be symmetrical, level, and not too prominent.

8. **Hind Legs**

- The hind legs propel the body forward. They must be strong and well made.
- From a side view, there should be plenty of length from the stifle to the hock and the muscle of the gaskin should be well-developed.
- The point of the hock should be directly below the point of the buttocks with a line down the back of the cannon bone dropping straight to the ground.

9. **Hocks**

- Hocks are the hardest worked joint. They should be large, wider when viewed from the side than the front, and deep.
- Avoid any animal with a swollen hock, avoid those lacking good flexion, and avoid those showing pain on flexion.
- Bent or sickle hocks are predisposed to strains and curbs.
- Cow hocks (toes out, hocks in) cause the limbs to move outward instead of forward. This results in strain.
- Bowed hocks (toes turned in and hocks out) cause the animal to twist the hock outward as the foot touches the ground and can cause strain.

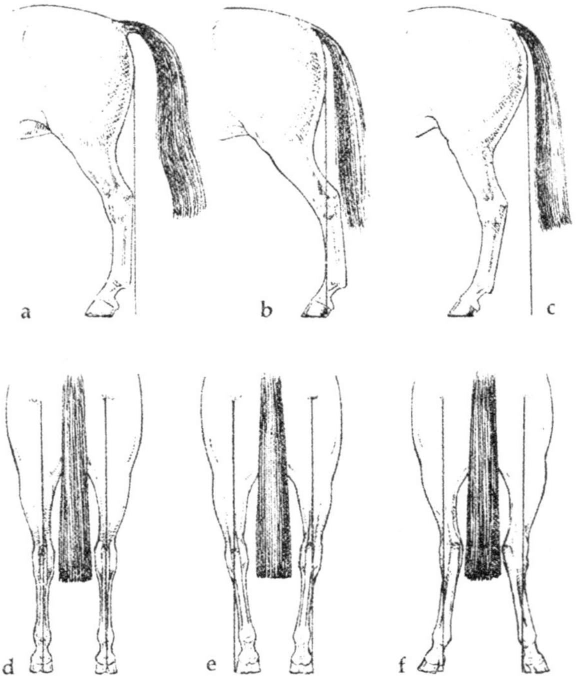

Hind legs. From side: **(a)** Correct. **(b)** Out behind. (Often hocks too straight). **(c)** Too far forward. (Often hocks too bent). From rear: **(d)** Correct. **(e)** Bowed hocks. **(f)** Cow hocks. *(Courtesy of The Manual of Horsemanship.)*

Action - It is important to see the animal move at a walk and trot from the side, behind, and the front.

1. **From Behind and in Front** - In walk and trot the animal should move freely and straight with the hind feet following the line of the forefeet. There should be no indication that the horse might strike his own legs by moving incorrectly.
2. **From the Side** - Look for lameness. The hind feet should over-stride the imprint of the forefeet.
3. **At a Circle** - Find hard ground and trot the animal in a small circle. Look for lameness or difficulty in turning.
4. **After Flexion of Leg Joints** - Flex each joint for several minutes; then trot the animal. If the animal flinches at the flexion or is lame after flexion when trotting out, there is some problem with that joint.

Attitude and Behavior

Horses, mules, and burros are very large animals and can easily injure the handler. For this reason, it is very important to select and use only those animals with the best attitude and disposition.

A problem animal can be described as one who behaves in an undesirable way because it refuses, or is unable to do what we desire. Animals that kick, strike, or buck should be avoided. Those with vices or bad habits such as cribbing, air sucking, or weaving may not lend themselves to be good work animals.

Selection of animals without undesirable traits is very important. Selling those with problem behavior may prevent severe injury. It is very important for the seller to warn potential buyers of problem animals unless they are tobe used for food. Do not breed animals with poor behavior as the offspring may carry the same trait.

Section 5

Equipment

Equipment in developing lands is very limited and may vary in different regions. Many harnesses are made from locally available materials.

According to field observations by Andre Bubear for WTFE (Working Together For Equines) "there is a strong regionalization with respect to harness style, material and hitching technique. There is no „right" or „perfect" harness for all the different kinds of working horses, donkeys, and mules. Time has refined basic harness designs to fit particular uses, animals, and vehicles/ implements."

"The problems relate more to a lack of knowing how to fit a harness properly and how to make repairs, or how to maintain it to prevent abuse of the animal." Nor do people understand what material should and should not be used in making harnesses. One must understand the basic needs forequipment and proper fitting to assist those in developing countries. Prevention of injuries, avoidance of painful working harnesses, and development of efficient gear should be your goal.

Halters - The most efficient way to secure a horse is by a halter and lead rope. The halter can be purchased or made of leather or rope (see diagrams). Never use a sliding loop to secure the head or neck as an injury may result.

Lead Ropes -

- Leads should be strong and not frayed. Avoid flat nylon as these cause injury to the hand and are difficult to untie.
- When tying an equine, avoid unsafe walls or rails which might give way should the horse pull back. Always tie to something very solid and strong.
- When tying, use a quick-release knot.

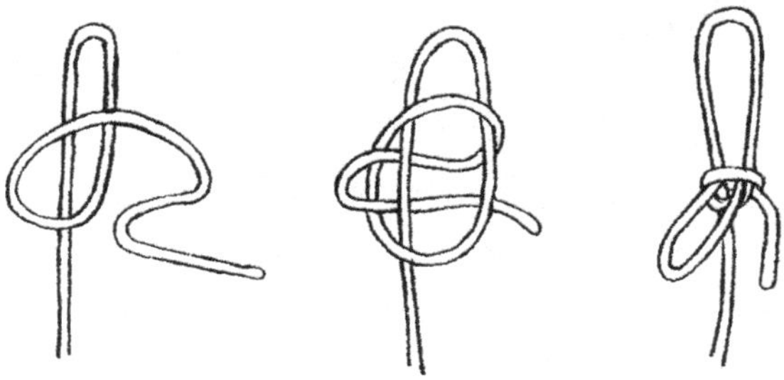

- Another good knot to know is a bowline

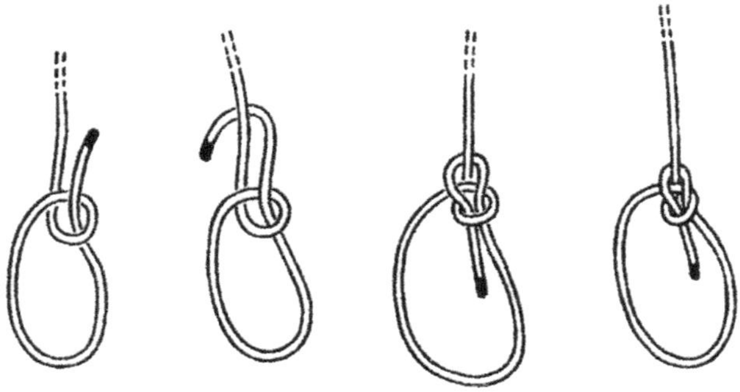

Bridle and Bitting

1. Rope Hackamore - This is a bitless bridle. There is not a mouthpiece but the hackamore acts by leverage on thenose, poll, and chin-groove. All of these pressure areas are sensitive and can be damaged by improper use.

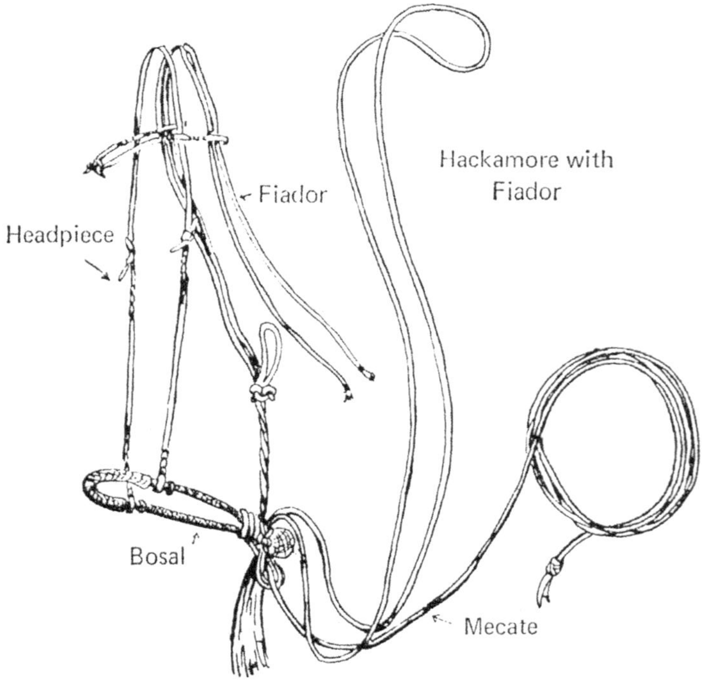

Complete hackamore with 22-foot hair mecate tied for reins and lead rope. A strong fiador and tie rope can be used as a halter when breaking colts.

2. Leather Bridle

- The headpiece and throat lash are made from the same piece of leather. They support the bit in the animal"s mouth. The throat lash when correctly fitted (not too tight) will prevent the bridle from being pushed forward over the ears.
- The brow band prevents the headpiece from slippingback.
- The cheek pieces are attached at one end to the bit and the other end to the headpiece.

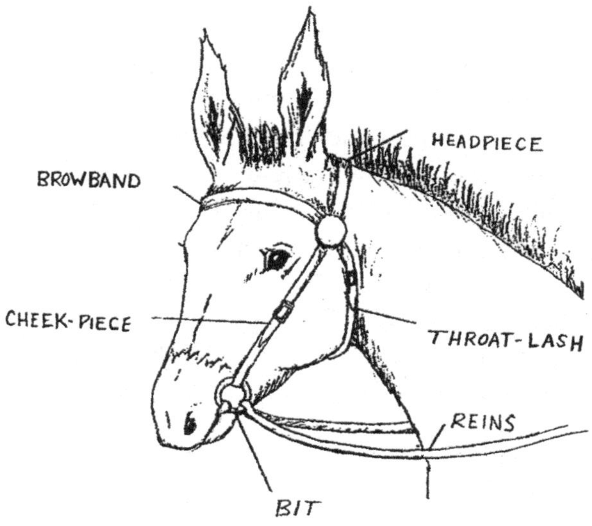

3. Bits - The bit is the part of the bridle that is fitted into the horse"s mouth over the tongue. There are three types of bits: the snaffle, the curb-bit, and the bar bit (pelham).

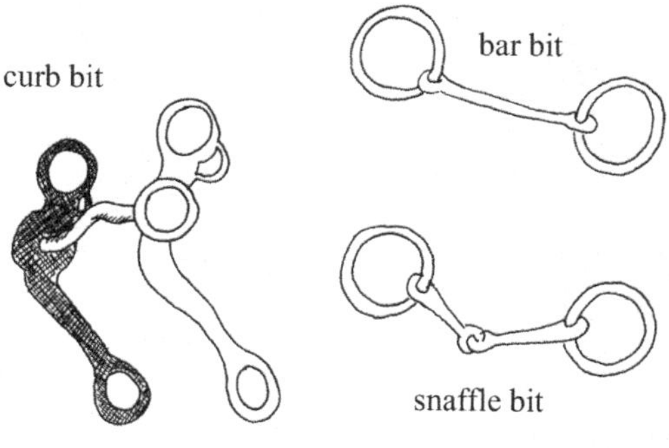

Fitting the Bit -

- A bit that is too narrow will pinch the corner of the mouth, while one that is too wide can move excessively from side to side causing soreness. If a jointed snaffle bit is too low in the mouth, the tongue will come over the bit.
- There should not be excessive pressure on the tongue. The tongue should lie comfortably below the mouthpiece.
- The mouthpiece shouldn"t be too thin as this is too severe on the bars of the mouth. A thick mouthpiece is kinder unless it is too large to be easily accommodated in the mouth.
- The correct width is when the mouth-piece extends 1/4 inch (5mm) past the side of the mouth.
- The correct height is such that the bit makes the animal smile but does not unduly wrinkle the corners of the mouth.
- No matter how well the bit fits, the animal will not be comfortable if there is any soreness in the mouth...be sure to look in the mouth.
- Any bit that does not fit may cause the horse to "pull" or to flee the pain. A tongue over the bit, a dry mouth, or an excessively frothy mouth are all indicators of pain.

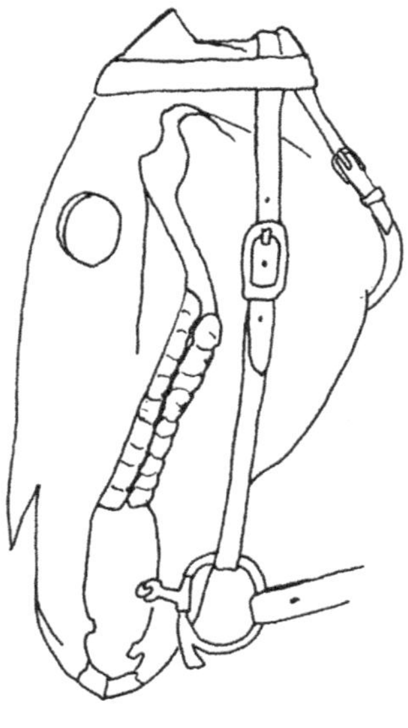

correct position

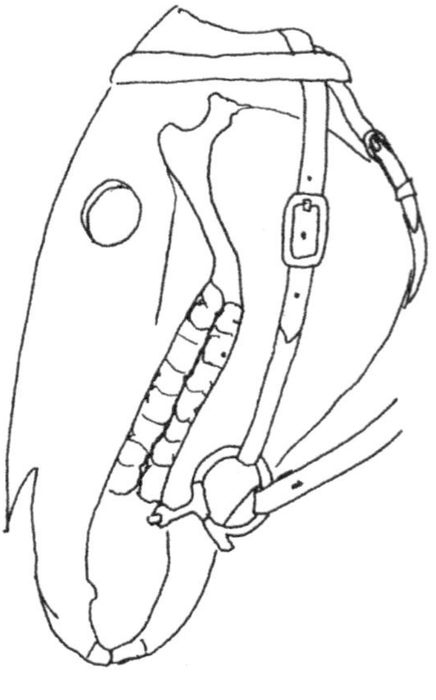

incorrect position

4. Reins

- They are made of varying widths and should fit the handlers hand. If they are too wide, they will be uncom- fortable to hold.
- Reins that are too narrow will slip through the hands too easily. The reins may be made of leather, rubber, nylon, or linen.

5. Packs and Pack Saddles - There are several types of pack saddles: the sawbuck, the Decker, the McClellan, and some homemade versions.

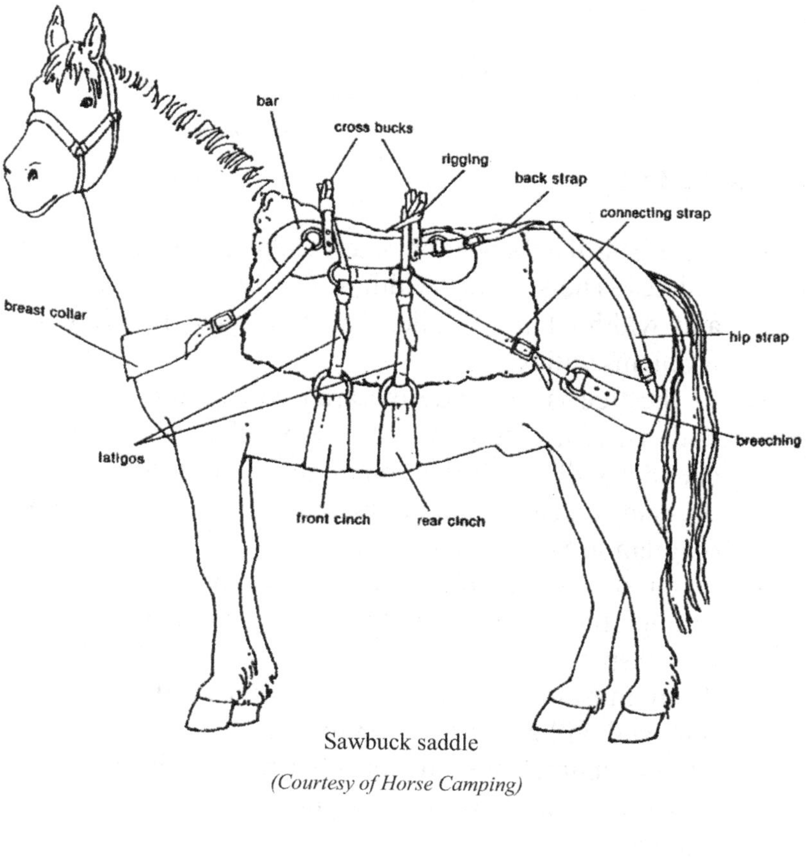

Sawbuck saddle

(Courtesy of Horse Camping)

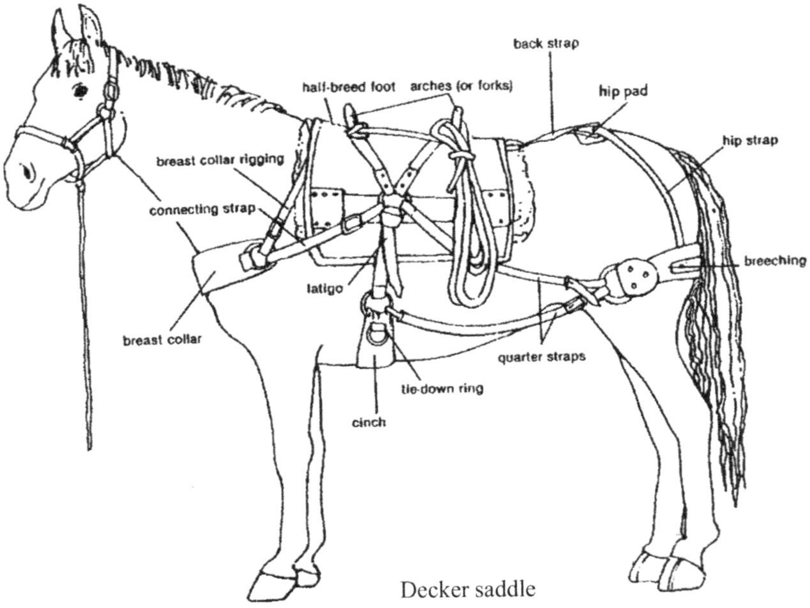

Decker saddle

(Courtesy of Horse Camping)

Saddle Fit

- The tree is the foundation of any saddle, either riding or pack saddles. The size and width of the tree determines the size and width of the saddle. The tree must fit the horse or problems will result.
- If the tree is too wide the arch will be too low over the withers. If it is too narrow, the panel on each side of the channel will pinch the spine and there will be excessive pressure under the points of the tree (see back & withers diagram in Section 4).
- You should be able to see a clear passage all the way through the channel over the horse"s spine with no contact on the spine.
- There must be no weight over the loins or the saddle is too long and will cause muscle problems.
- If the animal has any peculiarities (such as prominent withers), padding may be required to fit the saddle to the equine.
- The saddle should lie flat on the animal"s back. It should

not be tipped.

- The saddle must not hamper the movement of the shoulders. This may happen if it is too far forward.
- The saddle must not slip forward. Fat horses or those with flat withers allow this to happen. A crupper orbritchen will help prevent slippage from the rear and a breast collar from the front.
- The tree must not be twisted or broken as this will cause back damage.

Girths

- Girths are the means of securing the saddle on the back. They are also important for the comfort and efficiency of the horse.
- Materials most commonly used are leather, webbing, nylon, man-made fiber, and lamp wick.
- Be sure girths are not frayed or decaying. Avoid
- using girths with rough and broken parts.

Pads

- Pads are used under the saddle to protect sensitive areas from friction.
- Pads also temporarily minimize the effects of ill fitting saddles.
- Pads are made from sheepskin, cotton covered foam, or wool blankets. Wool blankets work the best, especially in humid areas.

Packing

- The key to packing is having a balanced pack on each side of the animal. The packs should also be at the same height on each side.
- The burden should be carried at the top of the body"s side or rib cage. Be careful not to set the pack too far forward to inhibit motion or too far back to cause muscle damage to the back.

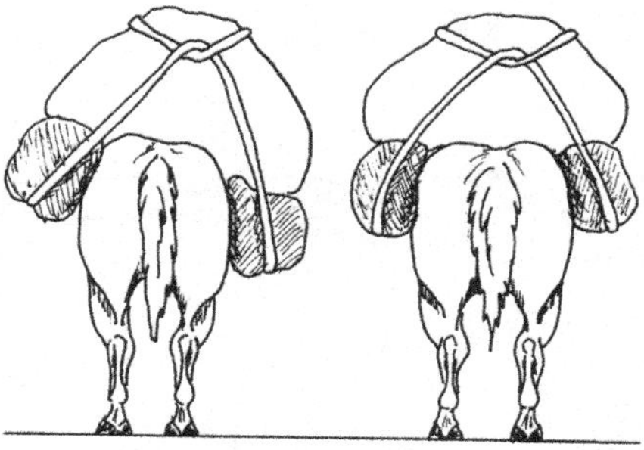

Balance your pack

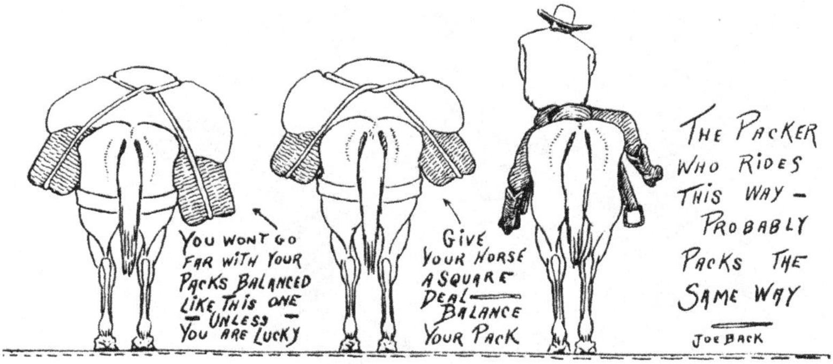

(Courtesy of Horses, Hitches & Rocky Trails.)

- If the pack is loaded too high it will slip soon.
- Packs loaded too low inhibit respiration and can cause body sores.
- Material can be secured with panniers made of wood, woven baskets, leather, nylon, or canvas.
- Care must be taken to see that the load fits correctly and the saddle or gear does not rub or pressure unnaturalareas. Rub sores or muscle problems can result from improperly set packs.

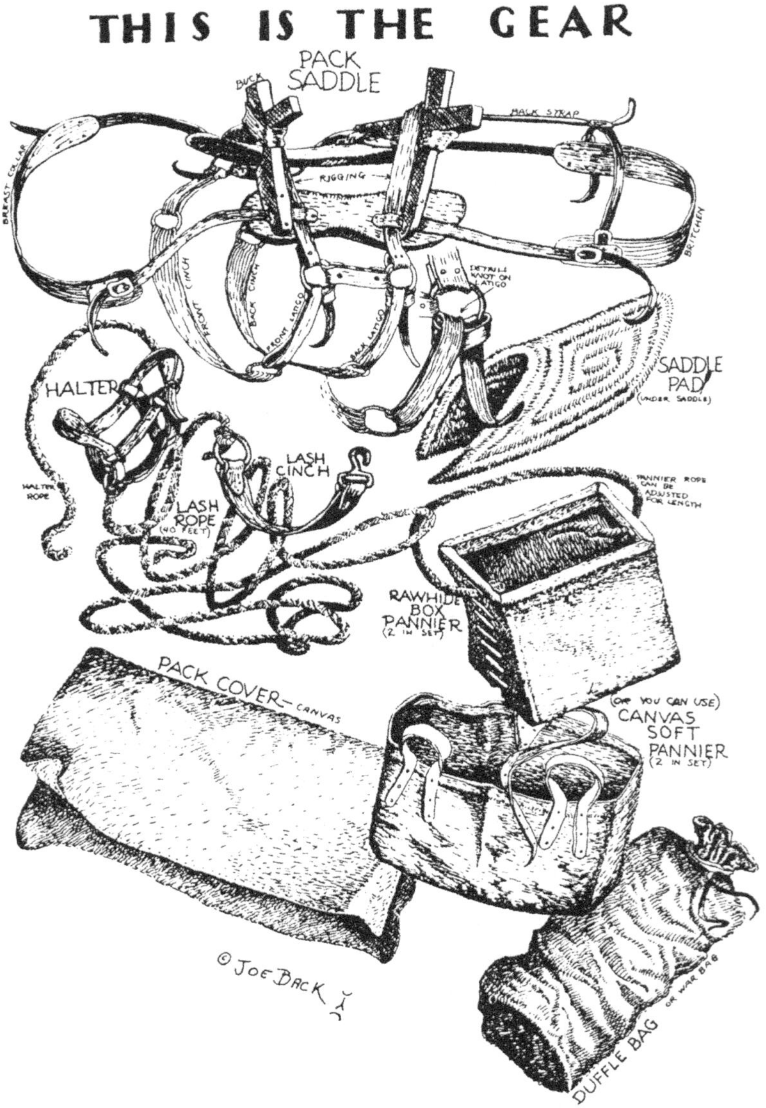

(Courtesy of Horses, Hitches & Rocky Trails.)

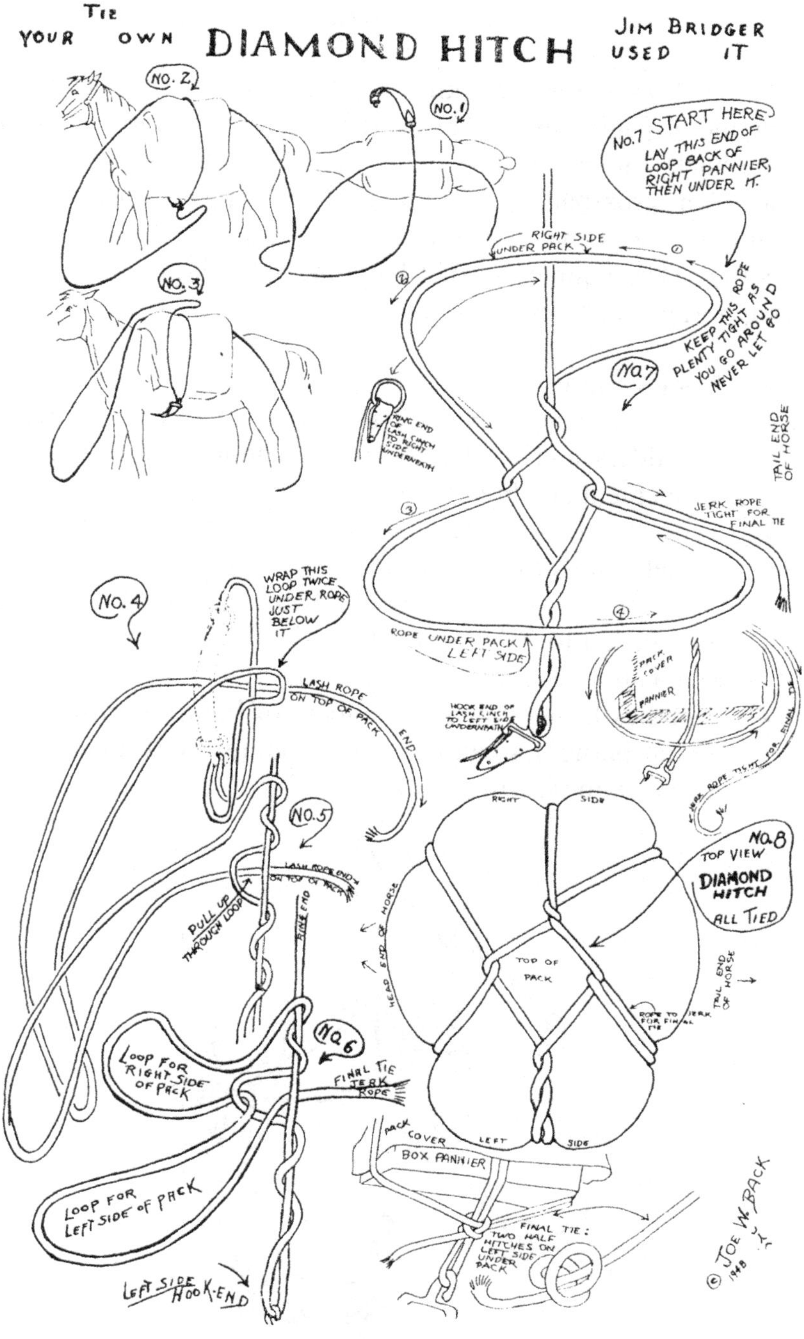

(Courtesy of Horses, Hitches & Rocky Trails.)

Draft Harnesses

The tack used for a workhorse differs greatly from that of a saddle or pack animal. The workhorse doesn"t carry loads on the back, rather the load is attached behind. The ox yoke, a wooden crossbar with two U-shaped pieces around the animal"s neck that rests on the withers, does not work for the equine. The equine doesn"t have the same high shoulders to support the yoke that an ox does.

1. Neck Collar

- This is a rigid, padded form that sits on the animal"s shoulders and circles the neck without interfering with movement and breathing.
- The correct size and shape of the collar is a must. If the collar is uncomfortable the animal will be unable to work well and sores can develop. Pads and adjustments may be needed to help adjust a poorly fitted collar if another one is not available.
- Studies have shown that a team of horses wearing neck collar harnesses can pull up to five times as much as the same team wearing a yoke.

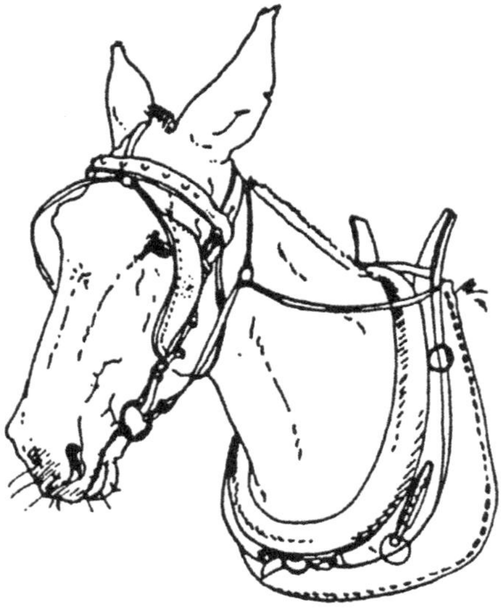

Single Wagon Harness

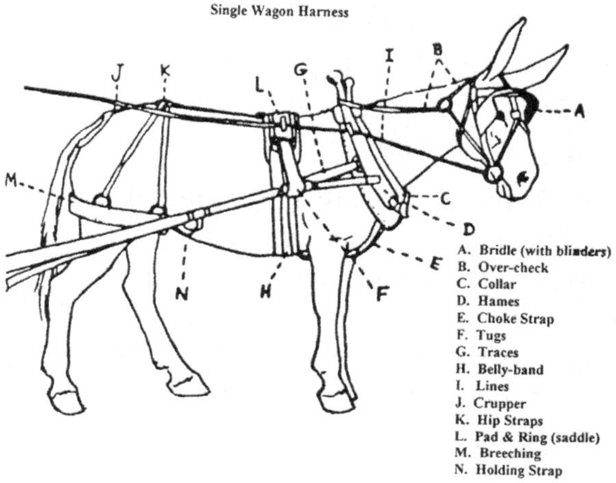

Single Plow Harness

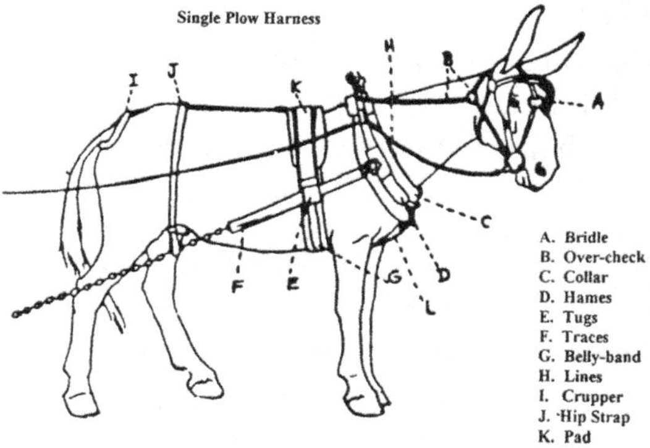

2. Hames

- Hames are two rigid curved pieces of wood or metal that form a frame around the collar.
- The hames are the most important pieces of harness besides the collar.
- They must lie full length along the collar and are fastened to it at the top and bottom by hooks called draft hooks. The reins go through rings on the hames known as terrets.
- Hames that are too long or too short reduce the animal"s pulling strength and can lead to tired or sore muscles.
- If there is an accident and the animal falls down or is choking, the hames must be removed quickly to take pressure off the collar and the throat.

- When the hame straps are cut, the entire harness will slip off the draft animal. Many horseman carry knives for just this purpose.

3. **Back Pad**

- The back pad goes across the animal"s back where the saddle fits on a riding horse.
- The pad is a narrow piece of leather that helps support the traces. The reins go through terrets here also.
- The back pad is held in position by the belly strap, the back band, and sometimes a crupper (a leather strap that passes under the animal"s tail).

4. **Breeching**

- The breeching is a series of straps that go behind the horse"s hindquarters
- Breeching helps the horse to stop or back up the load.

5. **Traces**

- Traces are long leather straps or chains that attach the work animal to its load.
- The traces run from the hames, along the animal"s sides, to the singletree behind the hind legs.

6. **Singletree or Whiffletree**

- This is a horizontal bar behind the rear legs that is attached to the load by a center link.
- The Singletree is made of wood in case the horse kicks, and helps equalize the pull of the traces.

7. **Reins**

- The reins run from the bit through terrets in the hames and back pad leathers.

The harness described is the best to maximize performance and to prevent muscles problems or sores. In developing countries all the parts may not be available. Most people make do with items available. If problems occur, careful observation with trial and error adjustments are needed to solve the problem and prevent more severe problems from occurring.

8. The Hitch

- Different formations can be used depending on the work needed to be accomplished. If three animals are needed they can be harnessed in tandem fashion, three abreast, or in a unicorn formation.
- The lead horse is the one at the head of the hitch. This animal must have quick action because it has the farthest to go on turns.
- The wheel horse is closest to the wagon. This is the hardest job with most of the pulling coming from this animal on turns. Strength and endurance are needed for this animal.

9. Charms

In developing countries some harnesses are decorated with charms and symbols to protect the owner and his property from evil spirits. These charms are often bright objects attached to the headpiece of the bridle. The "sun flashes" are believed by some to reflect the evil eye of a spirit away from the horse. Jesus Christ replaces all the need for such charms and this is an excellent opportunity to witness to those who use these charms..

Breast Collar Harness

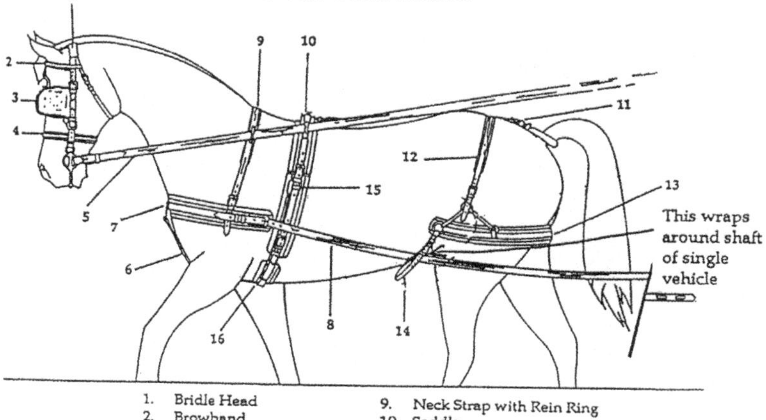

1. Bridle Head
2. Browband
3. Round Eye Winkers
4. Noseband
5. Reins
6. False Martingale
7. Breastplate
8. Traces, Slot Ends
9. Neck Strap with Rein Ring
10. Saddle
11. Crupper
12. Hip Strap
13. Breeching
14. Breeching Straps
15. Shaft Loops
16. False Belly Band

Section 6

Nutrition

In most developing countries the equine is not used for food consumption. Unless the owner understands that good feeding practices are needed to produce hard work, the horse will often be fed very poor quality feed in relation to other livestock. Type of work, type of feed available, and specific needs will vary but there are basic fundamentals that can be applied to most situations.

Feed Utilization - The equine digestive tract is proportionately smaller that that of the cow. Horses, mules, and burros must receive a diet with a high level of nutrients. In these animals protein digestion begins in the stomach. In the small intestine, more protein is digested, carbohydrates are digested, and vitamins and minerals are absorbed. Next in the cecum, roughages are broken down by bacterial action and digestion continues. Finally in the colon, water is absorbed and undigested residues are eliminated.

A simple diagram of the digestive system.

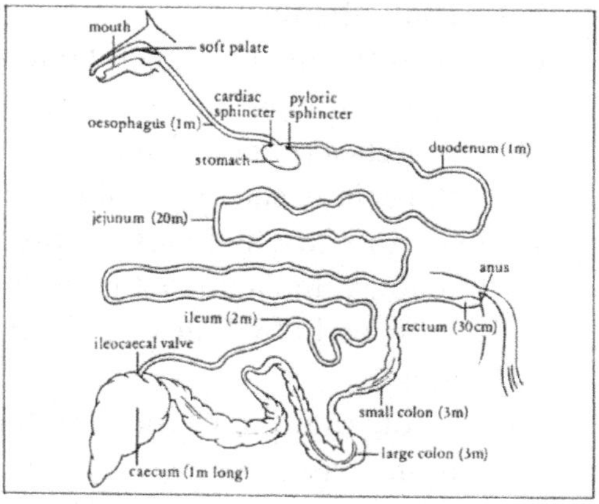

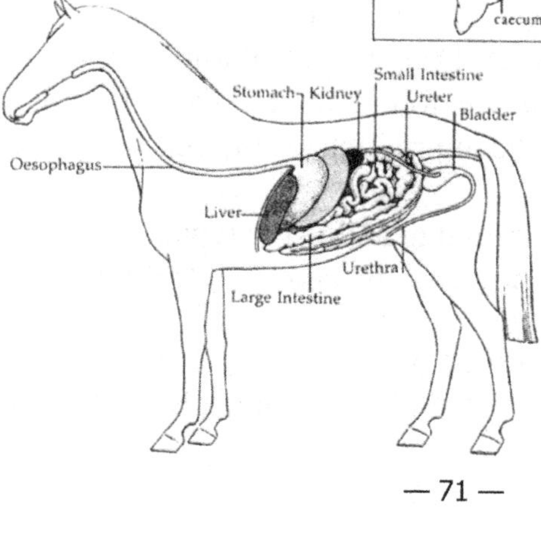

The digestive and urinary systems.
(Courtesy of The Manual of Horsmanship.)

Essential Nutrients

1. Water

- Water is the forgotten nutrient but is very important. The horse"s body contains 70-80% water. A loss of 1/10th of an animal"s body water will result in death.
- Water is required for normal utilization of all nutrients and plays a special role in regulating body temperature, joint lubrication, hearing, digestion, and waste elimination.
- Daily requirements vary from 5 gallons (20 liters) per day for idle activity (for 1000 lbs or 500 kg horse) to 12+ gallons (48 liters) per day for heavy work. A lactating mare needs 9-11 gallons (36-44 liters) of water per day.
- A supply of clean, fresh water should always be available. Horses should not be allowed to drink a heavy fill of water if they are hot from work or if they have been deprived of water for any length of time. Small drinks of water for several hours is preferred in these situations.

2. Carbohydrates

- This is the main source of energy.
- They are found in cellulose (grass and other forages), in starches (grains), and in sugar (molasses).
- Extra energy can be given in the form of grains if available.
- Lack of carbohydrates results in thin body condition, loss of energy and a horse that is cold.
- Excess carbohydrates results in fat animals.

3. Fats

- Carbohydrates are stored in the form of fat.
- Fat increases the energy level in feed, is necessary for absorption of vitamins A, D, and E, and it improves hair coat.
- Rich amounts of fat are found in milk-based feeds, rice bran, soy beans, soy bean products or linseed.

4. Proteins

- Proteins are for body building and tissue repair. A continuous supply is necessary for growth and maintenance throughout life.
- Proteins are made up of amino acids, half of which are produced in the body. The other half must be in the diet.

- Proteins are found in all food stuffs and the essential amino acids (those not produced in the body) are in cereals, especially oats.
- Peas, beans, lucerne and soya meal all have high levels of the essential amino acids but they should be fed with care and only to those animals doing hard work. Overload can cause severe indigestion and possibly death or founder.
- Dietary protein requirements range from 9% formaintenance to 18% for work, lactation, and growth.
- Lack of protein results in poor growth, poor performance, lack of appetite and condition, and poor milk production for mares.

5. Minerals

- Minerals are involved in a number of body functions necessary for effective metabolism.
- Minerals are building block for bones and teeth, and are involved in energy, vitamin, hormone, and protein utilization.
- Calcium and phosphorous combine with Vitamin D forbone development. The balance between these two minerals is very important. There should be 1.3% calcium to 1% phosphorous in the diet. Imbalances of these elements can cause crooked legs in growing equines. Cereal grains are very high in phosphorous. Alfalfa, limestone and bone meal supply extra calcium.
- Sodium and chloride are supplied as common salt. They are essential for regulation of body fluids. Animals at hardwork will need two to four ounces per day for good function.
- Potassium is very important for animals at hard work. It can be found in low herbal growth, tree bark, and cider vinegar.
- Selenium along with Vitamin E is important for good muscle function. Some areas can be high in selenium in the soil and other areas are very low. Check with authori- ties in your area to determine the local selenium levels. White muscle disease results from the lack of selenium andis treated with injections (E-SE...vitamin E & selenium) and feed supplements.

6. **Vitamins**

- Green forages and sunshine assure the equine of adequate Vitamins A, D, E, and K plus B-complex. The horse is able to synthesize most of the vitamins needed. Sick, debilitated, or anemic animals may need extra vitamins and minerals for proper healing and body rebuilding. Most cattle vitamins can be given to horses, but make certain the supplement contains no ionophores like Bovatec or Rumensin as they are toxic to horses.

Feeding Practices

1. **Water**

- Have clean, fresh water available at all times.
- With heavy work, additional water will be needed.
- Avoid stagnant water with potential toxic algae blooms or toxins from waste.

2. **Feed Often**

- A horse"s stomach is relatively small – 6.5 to 11 liters.
- By the time feed reaches the horse"s stomach the volume will have doubled due to saliva and digestive juices. The stomach is most efficient when about 2/3 full.
- Food passes through the digestive system in rhythmical muscular contractions (peristalsis). This actions means it is best to have a constant feed supply for the equine.

3. **Feed According to Work, Temperament, and Condition**

- Working horses should have the energy level of their diet (grains and sugars) decreased if they are not at work to prevent azoturia (muscle cramps and "tying up").
- Overweight horses need less feed and can be put in a dry lot for part of the day so they will not founder.
- If grain makes an animal excitable, feed roughage only.
- If the equine is thin parasites should be ruled out. Energy level of the feed may need to be increased to keep thebody in good condition.
- Young horses and broodmares need extra vitamins, minerals, and additional protein for good growth. Old horses do not chew and digest feed as well as others.These animals do better with additional grains and

- processed feeds if available. It may be more economical to cull old animals than to try and keep them in condition if feed is in short supply.

4. Make any feed Changes Gradually

- Sudden feed changes produce gastrointestinal upsets and colic.
- Gradually increase new feed and decrease old feed to prevent this GI upset.

5. Feed Clean Quality Feeds

- Avoid moldy or dusty feeds. These can cause colic and respiratory problems.
- The horse was meant to graze and chew 80% of the time and sleep or rest 20% of the time. Confined horses withoutgood bulk in their diet tend to chew fences and wind suck.

Types of Feeds

1. Pastures

- Good pastures are essential for successful feeding of horses, mules, and burros. Pastures should be well drained and supply adequate shade and water.
- Temporary pastures grown in a crop rotation are often preferable to permanent pastures that are heavily infested with parasites.
- Lush green alfalfa and similar feeds can cause excess gas production in the horse and resultant colic. Care should be exercised when there is access to these type of feeds. These are better utilized when they become dry or after a killing frost.
- Fescue grass pastures can cause problems in mares. Abortions, stillbirths, and poor milk production may be caused from a fungal toxin on the mature growth of the grass.
- Natural grasses in primitive areas will vary in their nutrient value depending on the time of year and the quality of the grass. Equines under heavy work may need extrasupplement of grains to be efficient.
- Poisonous weeds can be a problem in some areas espe- cially if the area is dry or overgrazed. Check with local authorities about poisonous plants to avoid.

2. **Hays**

- Oat straw is often fed in developing lands to horses, mules, and burros. It can safely provide up to one third of the roughage fed and is better than very poor hay but its nutritional value is low. Barley and wheat straws are very poor substitutes. Beards (head of plant) can cause severe mouth problems as they get stuck in the tissues.
- Grass hays are often fed to horses and work very well except for those in hard work, pregnant, lactating, or growing.
- Legume hays are higher in protein and minerals than grasses or oat straw.
- Hay should be high quality and free from dust or mold to prevent breathing problems or colic. Dusty hay can be soaked with water to prevent respiratory problems.
- Silage is not suitable feed for horses as there is a high risk of contamination by harmful bacteria, which can be fatal.
- Blister Beetles are present in some parts of the world. They can contaminate the hay and release a caustic chemical called canthardin. A few beetles, when swallowed, can irritate or destroy the lining of the intestinal tract. Symptoms range from colic and diarrhea to kidney failure, shock, and death.

3. **Grains**

- Oats work very well for horses because they are high in fiber, have a good protein content and are easy to feed.
- Corn, barley, milo, and wheat can be substituted in the grain ration. These grains must be cracked or rolled and they are „hotter", supplying more energy. Due to their highenergy, extreme care must be taken in feeding corn, wheat,or milo so the horse doesn"t founder. Fresh cut wheat is very dangerous.
- Grains are of particular value for animals in heavy work, pregnancy, lactation, or during growth.
- Bran is a byproduct of wheat milling. It is added to provide bulk in the diet and assist digestion. Excess bran can cause phosphorous problems.

4. **Salt & Minerals**

- This is very essential to the equine diet and more so with animals in heavy work. Salt is best made available in a salt lick, which enables the equine to consume the amount needed.

- Grain rations are marginal in calcium and it is advisable to feed a mixture of ground limestone and bone meal, dicalcium phosphate or defluorinated rock phosphate. A good commercial salt will suffice.
- Horses will normally eat 2 ounces of salt per day, depending on the exercise level.

Feeding Management

1. A regular feeding schedule is very important for equines.
2. A digestive system full of hay can be a hindrance to working horses so many horseman prefer to feed two-thirds of the hay at night when the animal has plenty of timefor eating.
3. Do not feed hay on the ground if the soil is sandy or the horse eats dirt. Ingested sand accumulates in the GI tract and çan result in colic or death. Massive doses of bran, mineral oil, or metamucil (psyllium) will help eliminate sand in some cases. Wood feed troughs are better for this situation.
4. Be sure all troughs and water receptacles are kept very clean to avoid illness.
5. Do not feed working animals big meals on days they are not working. Cut the feed intake proportionally to avoid azoturia (tying up).
6. Animals that are at work need more concentrates as the work load increases if they are too work at their maximum efficiency (see chart on following page).

7. **Feeding the Pregnant Mare**

- Most of the fetal growth occurs during the last three to four months of pregnancy. Mares need to be in good con- dition during this time and need good levels of feed to maintain this condition and the growth of the fetus.
- Calcium and phosphorous are very important during the last period of gestation. Legumes are higher in calcium and grains normally have good levels of phosphorous. A

ratio of 1.1 to 1 of calcium to phosphorous is very important.

- If the area is low in selenium, zinc, or copper, supplementation of these elements is needed. They can be in the form of feeds rich in these elements or via extra mineral supplementation. Levels of supplementation would be 1 mg per day of selenium.

8. Feeding the Foal

- Foals will normally start eating small amounts of grain when they are about one month of age. Protein requirements for growth is high and milk may fall short during this time.

DAILY NUTRIENT REQUIREMENTS OF HORSES

Body wt.	Total Protein	Digestible Protein	TDN	Ca	P	Vit. A 1000 IU
Lb.	*Lb.*	*Lb.*	*Lb.*	*Gm.*	*Gm.*	
Growing Horses 800 lb. Mature Weight						
200	1.10	0.75	4.20	14	11	1.7
400	0.90	0.62	5.90	17	13	3.3
600	0.80	0.60	6.50	13	13	5.0
800	0.70	0.50	5.80	9	9	6.7
Growing Horses 1,000 lb. Mature Weight						
200	1.20	0.84	4.20	16	11	1.7
400	1.10	0.75	6.20	15	12	3.3
600	1.00	0.67	7.10	14	12	5.0
800	.90	0.66	7.70	13	12	6.7
1,000	.90	0.60	6.80	11	11	8.3
Growing Horses 1,200 lb. Mature Weight						
200	1.80	1.10	4.70	19	16	1.7
400	1.40	1.00	7.00	18	17	3.3
600	1.20	0.85	8.20	18	17	5.0
800	1.10	0.75	8.40	18	17	6.7
1,000	1.00	0.70	8.40	12	12	8.3
1,200	1.00	0.68	7.80	12	12	10.0
Mature Horses at Light Work						
600	0.57	0.42	7.00	9	9	5.0
800	0.73	0.50	8.60	10	10	6.7
1,000	0.86	0.60	10.20	12	12	8.3
1,200	1.00	0.68	11.70	14	14	10.0
Mature Horses at Medium Work						
600	0.57	0.42	6.00	10	10	5.0
800	0.73	0.50	10.10	12	12	6.7
1,000	0.86	0.60	11.90	14	14	8.3
1,200	1.00	0.68	13.66	16	16	10.0
Mature Mares, Last Quarter of Pregnancy						
600	0.80	0.55	5.00	12	11	14.0
800	0.97	0.66	6.10	14	13	18.7
1,000	1.12	0.80	7.30	16	15	23.3
1,200	1.30	0.92	8.40	18	17	28.0
Mature Mares, Lactation Peak						
600	1.95	1.40	11.00	23	18	14.0
800	2.33	1.63	13.00	27	22	18.7
1,000	2.55	1.78	14.40	30	24	23.3
1,200	2.85	2.00	15.90	34	27	28.0

These recommended allowances are taken from the National Research Council's publication on Nutrient Requirements of Horses.

- A creep feeder, allowing the foal to enter a feeding area while restricting access to the bigger horses, works very well for early supplementation.
- A good grain ration fed at the rate of 225 to 340 gm per 45 kg of body weight is an excellent creep feed. Legume hay is also needed for good calcium levels and growth.
- If crooked limbs occur, the diet may not be balanced in calcium-phosphorous or may not have the needed levels of selenium, copper, and zinc.
- Weaning should be possible at four months of age, especially if the mare is thin. Foals should definitely be weaned by six months of age and should be on good feeding programs.

SAMPLE RATIONS

Creep feed for suckling foals and weanlings

	A	B	C
Rolled oats	70.0	45.0	45.0
Cracked corn	---	18.5	23.0
Soybean oil meal	---	15.0	20.0
Linseed meal	11.0	---	---
Wheat bran	10.0	---	---
Dried whey	---	10.0	---
Alfalfa meal	---	5.0	5.0
Molasses	7.0	5.0	5.0
Limestone	---	0.5	0.5
Steamed bone meal	1.0	---	---
Defluorinated phosphate	---	0.5	1.0
Trace mineralized salt	1.0	0.5	0.5
	100.0	100.0	100.0

Yearlings, Lactating Mares, and Stallions

	A	B	C
Oats	54.0	50.0	60.0
Cracked corn	10.0	14.0	18.0
Rolled barley	20.0	---	---
Wheat bran	10.0	10.0	---
Soybean oil meal	---	---	10.0
Alfalfa meal	---	5.0	5.0
Linseed meal	5.0	7.5	---
Molasses	---	12.5	5.0
Steamed bone meal	0.5	0.5	---
Limestone	---	---	0.5
Defluorinated phosphate	---	---	0.5
Trace mineralized salt	0.5	0.5	1.0
	100.0	100.0	100.0

Section 7

Reproduction

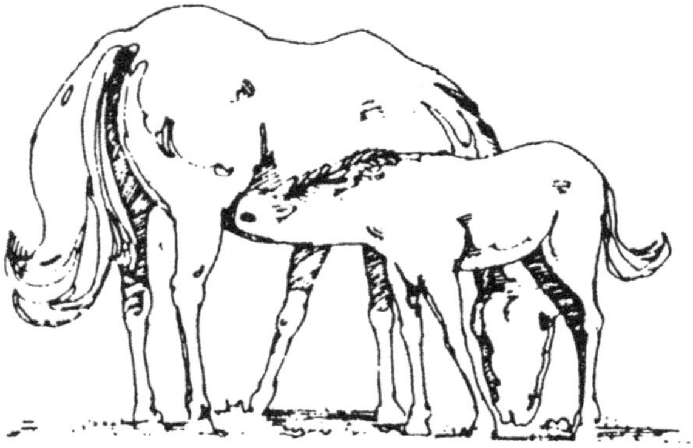

Basics Of Reproduction

1. Selection

- **Type -** Before breeding animals be sure you think about the desired end result, then mate the animals that will best provide that end result. Do not breed animals just because they are available. Pick only the best mare or jenny and breed only to the best stud or jack for the desired purposes. For example, if breeding draft animals that pull plows, do not breed small boned animals lacking muscle needed for pulling big loads. Also refer to Section 5 for selection information.
- **Behavior & Attitude** - Good and bad characteristics are often passed along in offspring. It is best not to breed animals difficult to handle, those that are mean, or those with bad habits.
- **Inbreeding** - Inbreeding is the mating of related animals. Avoid breeding stallion-to-daughter or other closely related animals. Inbreeding can cause the production of genetic defects or unwanted traits.

2. Definition of Breeding Terms

- **Season of Breeding -** The horse and donkey are seasonal breeders. These animals cycle and breed when the daylight hours are increasing which is May-June in the Northern Hemisphere and November-December south of the equator.
- **Heat (Estrus) -** This is a six to seven day period when the mare or jenny accepts the stud for breeding.
- **Heat Cycle -** The interval between heat or estrus periods. In the horse and donkey this is 19-21 days.
- **Foal Heat -** A first heat or estrus in the mare, 5 to 10 days post-foaling, which only lasts a few days. Foal heat has a lower fertility rate as the mare is cleaning out the uterus after foaling.

- **Ovulation** - This is the release of the egg from the ovary into the oviduct where it is fertilized by the stud's sperm. Ovulation generally occurs near the end of the estrus period.
- **Conception** - Fertilization of the ova or egg by sperm following mating.
- **Gestation** - Length of pregnancy.
- **Foaling** - Act of giving birth.
- **Dystocia** - Difficult birth.
- **Neonate** - Term for newborn the first two weeks of life.
- **Infertility** - Reduced or inability to become pregnant.
- **Ovary** – Where eggs are produced.
- **Oviduct** – Tube leading from ovary to uterus.
- **Clitoris** – Finger-like projection just inside the vulva of a mare.

3. The Stallion or Jack

- A breeding stud should have two, even sized, large testicles. Sperm are produced in the testicles, and large testicles convey more sperm and better fertility. An abnormal sized testicle can indicate disease or damage to the testicle.
- High environmental temperatures can influence the quality and quantity of sperm as can fever or infection in the stud.
- A good fertile stud receiving good quality nutrition should be able to breed up to 40 mares per year in natural breedings.
- Most studs can successfully breed a mare or jenny by the time they are 6-8 months of age. Animals this age should either be castrated or separated from mares so indiscriminate breeding does not occur.
- Do not breed animals with a retained testicle (cryptorchid), parrot mouth, crooked legs or other genetic defects.
- Studs breeding many mares or jennys need to be kept on a good feeding program to avoid weight loss and subsequent loss of fertility. Deworming is also very important. Do not breed before 3 years of age.

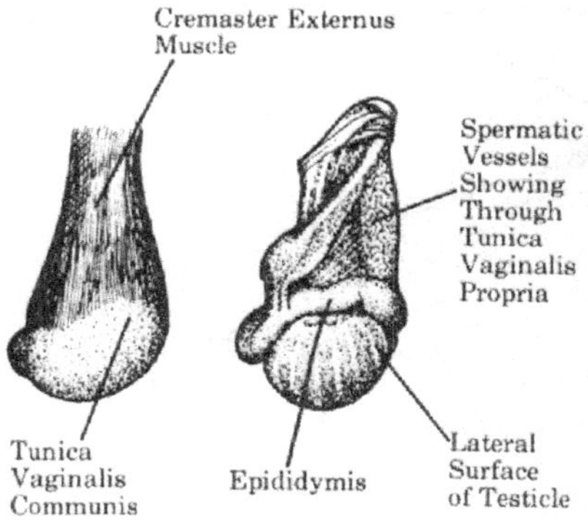

(Courtesy of The Illustrated Veterinary Encyclopedia.)

4. The Mare or Jenny

- The mare or jenny should not be bred before it has reached it's adult size. It is best to wait until they are at least three years old before breeding starts. Breeding young animals can stunt their growth and result in foaling complications or even death to the mare and foal.
- It is good to maintain breeding, pregnancy, and foaling records on mares so problems and trends can be spotted with infertility. Mares show heat by squatting, urinating, and everting the clitorus (winking). Most mares in heat will allow a stallion to tease and mount her. Mares not in heat will strike with the front foot, kick with the back feet, and squeal at the stallion. Some mares will be protective of foals and not show visual signs of heat when they havea foal at their side.
- Breeding mares must be kept in optimum shape. Overweight mares do not get pregnant easily. Very thin mares may not have normal cycles and may not carry afoal to full term. Deworming is also very important for these mares.

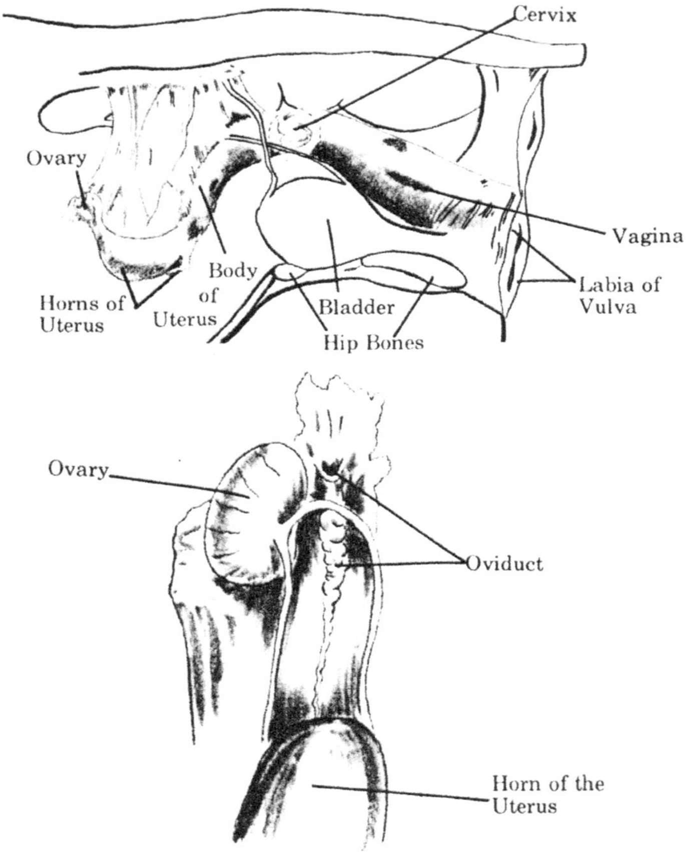

(Top): This side view shows the major components of the female reproductive system. The cervix is normally enclosed by the vagina. *(Bottom):* The ovum originates in the ovary, then travels through the oviduct to the horn of the uterus. *(Courtesy of The Illustrated Veterinary Encyclopedia.)*

Breeding

1. Methods of Breeding

- **Pasture Breeding** - This is the most common method of breeding. The stud or jack is maintained in a pasture environment with the mares and he breeds the mares as they come into estrus. An experienced stud will tease a mare and breed her when she is in full estrus. Studs can be injured from kicks by non-cooperating mares with this method. Breeding records are also more difficult to maintain because you will often not know when the last breeding occurred.

- **Hand Breeding** - This method involves a handler for the stud and the mare. If the mare has a foal at her side, the foal is often positioned with another handler behind a 5 foot barrier just in front of the mare so it is not injured. The stallion handler then lets the stud tease the mare to see if she is in acceptable heat. If she is willing to stand for the stud and exhibit normal estrus signs, the stallion is allowed to mount and breed the mare.
- **Artificial Insemination** - This technique involves the use of an artificial vagina that is used to collect the semen from the stud when he mounts the mare. This is nota common method in developing countries due to the equipment and training needed to collect and handle the semen.

2. Preparation for Breeding:

- The stallion should be washed with mild hand soap and rinsed well while the penis is extended. This cleaning should be done at the beginning of each breeding season and when a new mare is hand bred. Do not over wash or use disinfectants because this can kill beneficial bacteria and allow disease-producing bacteria to populate the penis surface.
- The mare should have her tail wrapped with bandage material or a clean cloth to avoid allowing tail hairs to cut the penis of the stud. The mare should also be cleaned with a mild soap like Ivory dish soap and rinsed twice withclean water.
- The above procedures are not always practical with pasture breeding.
- Breeding Hobbles can help prevent injury to a stud or jack by an aggressive mare that kicks.

Breeding hopples are used to prevent a mare from kicking a stallion. These hopples are also useful to restrain both rear legs for rectal or vaginal operation or examination.

Form a rope collar like that used in a rope sling, or use a web collar. Pass both ends of the rope back between the front legs, then pass one rope around each rear leg above the hock from the lateral to the medial side.

(Courtesy of Restraint of Animals)

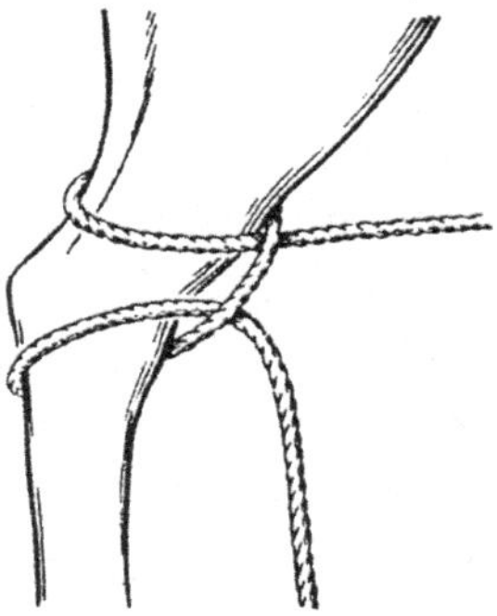

Carry the rope over the standing part, then back around the leg below the hock, from the medial to the lateral side, and pass the end under the rope as it crosses the medial side of the leg.

The ends of the rope can be drawn up and tied at the hock, as illustrated, or they can be taken forward and tied to the collar.

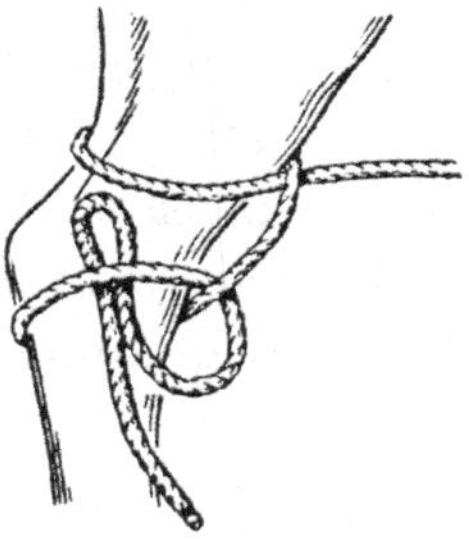

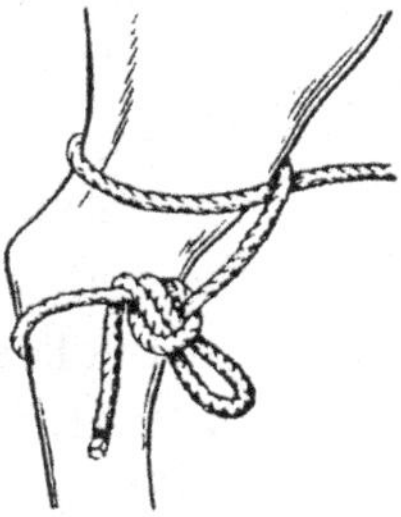

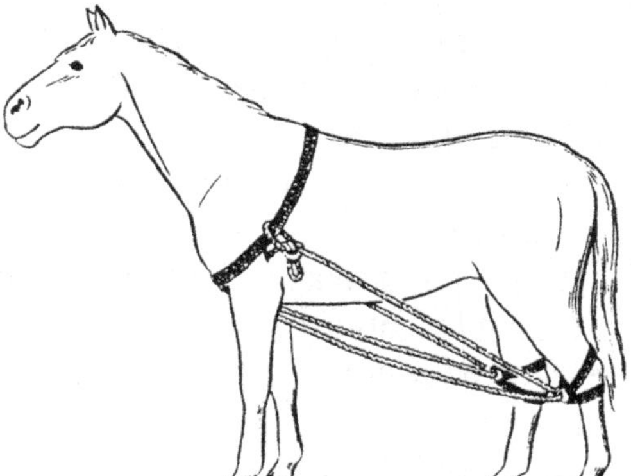

A special web collar may be used in place of the rope collar, and "two-way" web hock hopples may be used instead of the ropes around the hocks.

(Courtesy of Restraint of Animals)

3. Breeding Intervals

- Mares have the most fertile heat cycles during the period of year when days are the longest - increasing sunlight produces better ovary function. For a month or twoleading up to this optimum period, mares may show sporadic cycles or extended cycles but they are often not fertile nor ovulating during this time. The best months to breed a mare or jenny is 11 months or 345 days prior tothe desired birth of an offspring. Climate conditions or feed availability often dictate these prime foaling months.
- During the most fertile times, mares will cycle every 20 to 21 days. They show no heat for 14 days of this cycle with the remaining 6 to 7 days in estrus. Most mares ovulate near the end of the cycle. It is best to breed on the second day of the cycle and then breed every other day until the mare goes out of heat.
- If the mare is in heat for more than 9 or 10 days it is best to wait for a more normal cycle before continuing breeding. Mares over bred may develop infections and infertility.
- It is best not to breed during foal heats, but allow the mare to naturally clean her uterus.
- If a mare is in poor condition she should not be bred while she has a foal at her side.
- Mares that are not to be bred must be kept away from the stud in pasture breeding operations; either in another secure pasture or in a fenced enclosure.

4. Breeding Problems - Stud or Jack

Lack of libido or breeding desire can occur in youngstuds, in those previously injured during breeding, or it may be due to a hormone problem. Additionally, some studs may not like a certain mare or jenny and refuse to breed to them. If the animal is young, give it more time todevelop. If the low libido is after a breeding injury, you may have to work with the animal in a hand breeding situation using a very gentle, hot mare. Since survival of the fittest is a natural selection process, some low libido studs should be castrated so the problem is not perpetuated.

- Tail hair can cause cuts on the penis of the stud. Blood will kill equine sperm so breeding in these situations is useless. Allow the cut to heal. Use screw worm or fly repellent to avoid further problems from these externalparasites. Once the area is totally healed, breeding can commence.
- Kicks to the testicles or penis can cause severe damage. If such an accident occurs, sexually rest the stud until all swelling subsides. Some traumas may render the animal useless for breeding.
- Venereal infections caused by bacteria, viruses, or protozoa do occur and can render the stud infertile. If a stud has any discharge from the penis, swelling of thepenis or testicles, or is not getting mares pregnant, do not breed the animal. Have the affected animal examined bya veterinarian or castrate him.

5. Breeding Problems - Mare or Jenny

- Again, selection and use of the most fertile and prolific animals is of prime importance so undesirable traits are not passed on to successive offspring. If mares have fertility problems it is perhaps best not to breed these individuals in case this tendency is passed on.
- Genetic defects of the reproductive tract will make pregnancy impossible. Animals with genetic defects (see Section 5) should not be bred but culled or used only for work if they are fit.
- Mares that produce a discharge from the vulva or do not get pregnant may be infertile.
- Once mares reach the age of 10 to 12 their defense mechanisms of the uterus that naturally limit infection decline. Some of these mares can be treated so they can produce foals; however older mares that do not produce foals would be best culled and the financial resources invested into more productive areas than treatment.

- Poor condition in the mare, due to parasite load and/or poor nutrition, is the most likely cause of infertility. Get these conditions corrected well before breeding.

Gestation

1. **Diagnosis of Pregnancy**
   a. **Subjective history and signs:**
      - Serviced by a stud.
      - Cessation of estrous periods through 45 days after being bred.
      - Increased gentleness by the third month of pregnancy.
      - Enlargement of the abdomen by the fifth month of gestation.
      - Observation of fetal movements especially after drinking cold water by the 7th to 8th month.
      - Enlargement of the udder.
      - Relaxation of the pelvic ligaments by the 10th month

   b. **Rectal examination**
      - Rectal examination by a trained person can detect enlargement of the uterus by the 40th to 45th day of gestation.

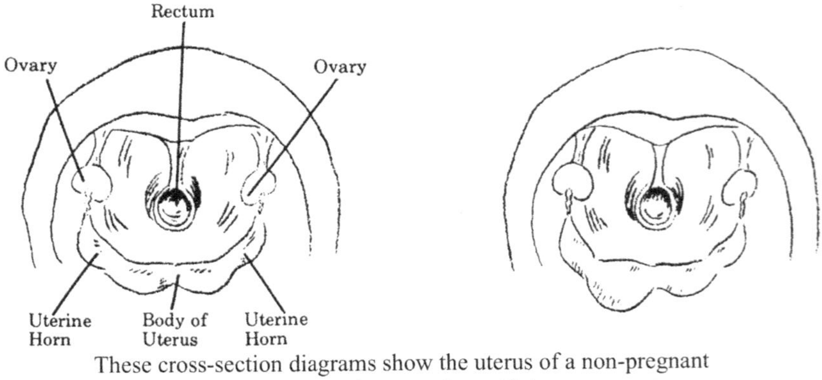

These cross-section diagrams show the uterus of a non-pregnant mare, and the uterus of a mare about 60 days pregnant.

(Courtesy of The Illustrated Veterinary Encyclopedia.)

2. **Length of Gestation**
   a. Horses - 327 to 357 days with the average being 342 days.
   b. Draft mares - slightly shorter, 330 to 340 days.
   d. Donkeys - 356 to 375 days.

3. **Mare Care During Gestation**

- **Feeding -** Two thirds of the fetal growth occurs in the last 3 months of pregnancy. A good hay or quality grass along with a mineral supplement is important during this growth period.
- **Exercise or Work -** Most pregnant animals in good body condition can be worked moderately until the 6th month of pregnancy.
- **Deworming -** Deworm the pregnant mare every 2 months during pregnancy to help carry a more healthy foal. Do not deworm the last 30 days of pregnancy. Avoid organic phosphate deworming products in pregnant mares.
- **Vaccinations -** If available, herpes virus vaccine (Rhinopneumonitis or Pneumobort-K) should be given on the 5th, 7th, and 9th month of pregnancy. Give a tetanus, influenza and encephalitis vaccination a month before foaling.
- **Hoof Care -** Keep the feet trimmed and in good condition during pregnancy.

4. **Abortion or Fetal Loss**

- Some mares will have twins and often lose both fetuses, or abort a large and small fetus in later pregnancy.
- Bacterial infection of the uterus will cause the largest percentage of abortions.
- Viral disease like the Herpes virus, Viral Arteritis, or Equine Infectious Anemia will cause abortion or stillborn foals.
- Congenital or genetic defects can cause fetal death.

- Fescue grass has a mold that will cause edema or fluid accumulation in the placenta causing abortion.
- Thin mares not receiving adequate feed can abort or produce weak foals that soon die.

Causes of abortion should be determined if economically feasible. Save the fetus and placenta, and have them examined along with the mare soon after the abortion by a veterinarian if possible. Many situations can be prevented or treated in future gestations. Cull mares that abort several times.

5. Foaling

Parturition (foaling) in the mare is a rapid and violent process. The incidence of dystocia in horses is about 4% which is low when compared to other species. However, when dystocia exists the life of the foal is usually in jeopardy and the health and reproductive function of the mare can be compromised.

a. Stages of Parturition

- **First Stage Labor -** This is a process in positioning of the fetus in the uterus so the feet and head are extended and the foal is upright. Some mares show minimal signs of discomfort while others show marked colic (abdominal discomfort). Two to four hours are required for completion of this stage, which includes dilation of the birth canal.

Characteristic stances of first stage labor

- **Second Stage Labor** - This stage is initiated by membrane rupture and release of watery fluid. The fetus enters the birth canal and abdominal straining begins. Most mares lay down for delivery of the fetus. This stage is short and is usually completed with a delivered foal in 20 to 30 minutes. Because the fetal membranes (placenta) separate from the uterus with the onset of this stage, a fetus not delivered quickly is deprived of oxygen.

The bubble-like amnion appears shortly after the mare "breaks water."

- **Third Stage Labor** - Following delivery of the foal, the uterus continues to contract to remove the placenta and decrease the uterine size. The placenta is usually removed in 30 minutes to 3 hours.

Placenta being delivered after birth of foal

b. Dystocia: difficult birth or delivery.

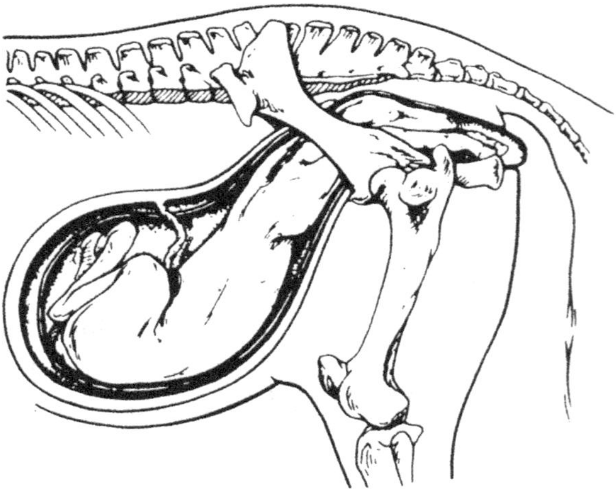

Normal presentation

(1)

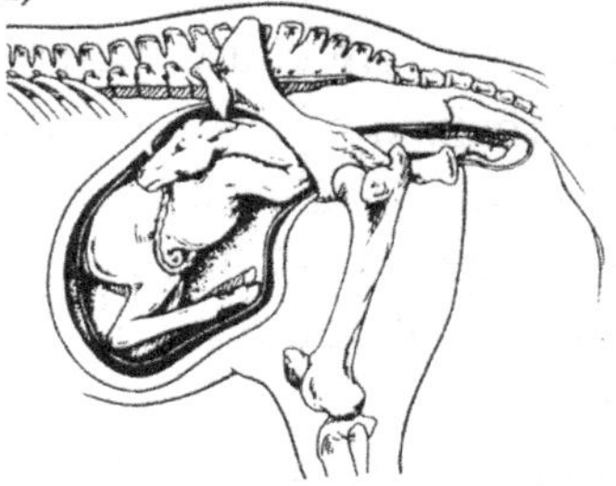

NO HEAD VISIBLE: The forelegs are emerging, but the head is pulled back and locked on the hips.

(2)

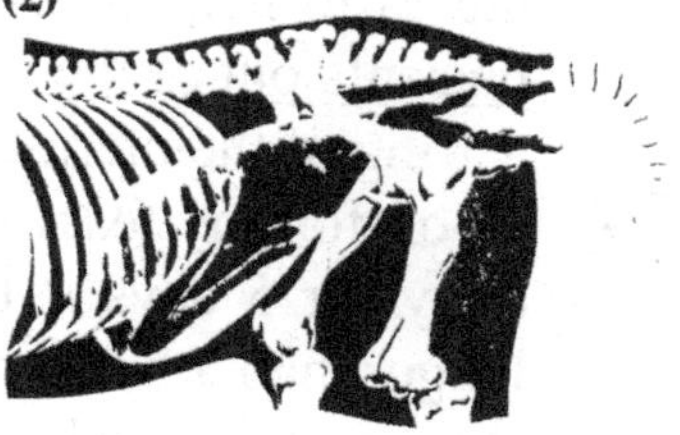

ONE FOOT, HEAD VISIBLE: The second forefoot is wedged in the roof of the vagina, where the next contraction could cause a perforation fatal to both mare and foal.

(3)

HEAD, NO LEGS OR FEET: Both forelegs are caught on the mare's pelvis and must be brought out. This is the worst possible position a foal can be in.

(1) You see two front legs up to the knees but no head, i.e., head back.
(2) You see only one front leg up to mid-cannon bone.
(3) You see a head but no legs or feet, i.e., both legs back.
(4) You see two feet facing up followed by the hocks. Normal breech.
(5) You see the foal's tail but no legs or head. Abnormal breech.

Other situations requiring immediate veterinary care:

(6) Foals too large to pass thru the birth canal
(7) Twins
(8) Premature, dead or deformed foals.

(4)

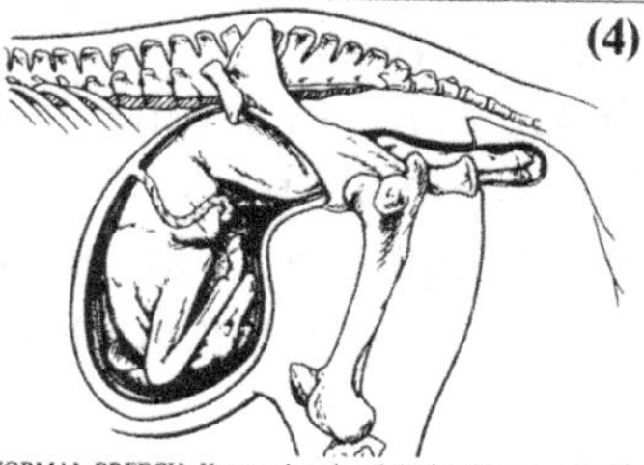

NORMAL BREECH: You see the soles of the feet facing up, but no head. If hocks follow instead of knees, a breech birth is certain. Delivery must be rapid because of the danger of suffocation should the umbilical cord be pinched. Once both hind feet are out as far as the hocks, the chances of a successful delivery are excellent, so long as the stifles do not bunch up at the rim of the pelvis.

(5)

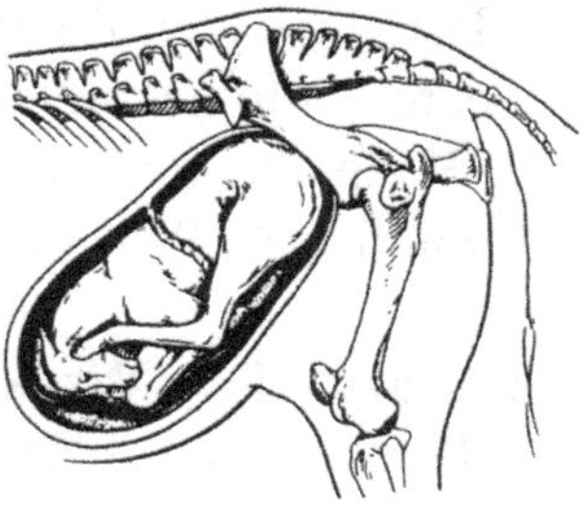

"Breech." Buttocks presented with hind legs retained.

c. Care of the Mare After Foaling

- Deworm the mare with Ivermectin (Equvalan) if possible the day of foaling.
- If the placenta does not pass in 3 hours, administer oxytocin and place a small weight on the exposed

uterus to help slowly expel it. Do not ever pull the placenta out of the uterus as this will cause bleeding, tearing of the uterus, and perhaps death of the mare.

- Look for any tears in the vagina and give Penicillin daily for 5 days if there are any tears or if it takes longer than 3 hours to expel the placenta.
- Be sure the mare has milk.
- Feed the mare extra so milk quantity is sufficient for the foal.

d. Care of the Neonate (first 2 weeks of life)

- The foal should stand and nurse within 3 hours of birth.
- If it is weak, milk the mare and tube-feed the foal or try to bottle-feed it, being very careful that it does not aspirate milk into the lungs with a poor suck reflex. Never force milk into the mouth of a foal without a good swallow reflex.
- The first 24 hours of milk (colostrum) from the mare are very important to the health of the foal. Be sure the foal nurses well during this period.
- Dip the navel of the foal in Betadine or Iodine. Take care not to get any of this liquid on the body.
- Keep the foal out of hot sunlight or cold, windy conditions. Provide shelter in cold or hot conditions.

e. Foal care - Two Weeks to Weaning

- Parasites are most damaging to foals because foals lack any natural immunity to the worms. Foals should be dewormed at 45 to 60 days of age and then every 60 days until they reach two years of age. The best products are Ivermectins (Eqvalan), given orally.
- Start vaccinations at 4 months of age with a combination containing tetanus, encephalitis, and influenza. If Rhinopneumonitis vaccine is available include it as well. Booster at 5 and 6 months of age.

- The hooves should be rasped at two to three months of age and then every 4 weeks. If the limbs are crooked, corrective trimming and rasping may help the problem. Consult with a veterinarian about methods of trimming for the different deformities.
- Weaning can be done at 6 months of age. Wean at times of least stress and after the foal has been dewormed and vaccinated. Often the foal will have to be separated from the mare for many months to prevent readoption by the mare.
- After weaning stress has subsided, castration can be performed. Horses bleed easily so it is best to have the proper equipment and a trained person perform the procedure. Very few males are of the quality that warrants saving as a stallion.

Section 8

Parasites

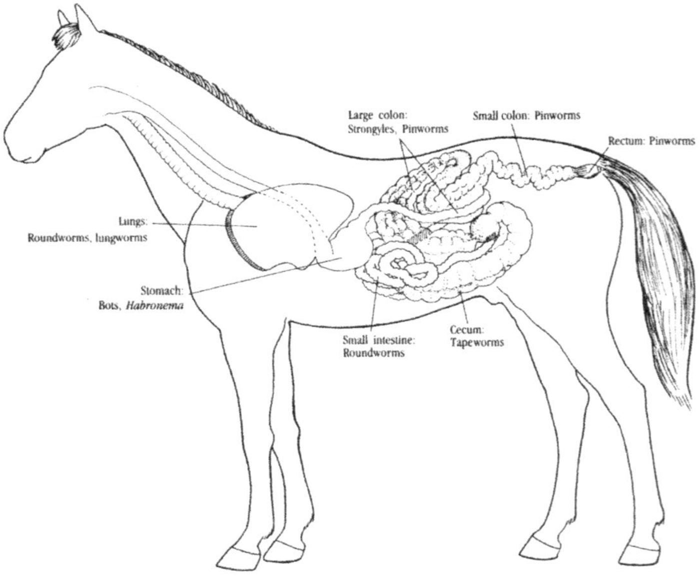

(Courtesy of Easy Health Care for Your Horse.)

Parasite control is one of the most important aspects of equine health. Internal and external parasites rob the horse of nutrition; they cause tissue damage, undue stress, and reduced immunity; and they can introduce disease. It is difficult if notimpossible to eliminate parasites but they must be controlled toan acceptable level for the equine to function at good health levels.

Glossary of Parasite Terms

Anthelmintic: Drug used to eliminate parasites from the host.
Ectoparasites/External Parasites: Parasites that live outside the animal and attack the skin and body openings.
Host: a human, animal, plant, or other organism in or on which another organism, especially a parasite, lives.
Internal Parasites: These parasites live inside the body of the host animal, also called endoparasites.
Intermediate Host: A host that is necessary for part of the development of an immature parasite.
Life Cycle: Stages of development the parasites goes through during its life.

Ectoparasites

Ectoparasites are divided into two main groups, arachnids and insects. The arachnid class includes ticks and mites. The insects include flies, mosquitoes, fleas, and lice.

1. **Ticks:** Ticks thrive on blood obtained from the host, causing devastating damage. They are divided into soft and hard ticks according to their structure. Ticks have a one host, two host or three host life cycle, as described later. Ticks occur worldwide but are especially trouble- some in tropical and subtropical countries where they are avery grave threat to the livestock industry. Ticks transmit disease and suck blood from the host. You will see ticks if they are a problem unless they are in the ears. If the tick is in the ears the horse will shake it's head often. Ticks are best handled by manually removing them. Grasp next to the skin with tweezers and pull straight out. Use tick sprays or dips to prevent reinfestation.

(Live on the outside of the animal)

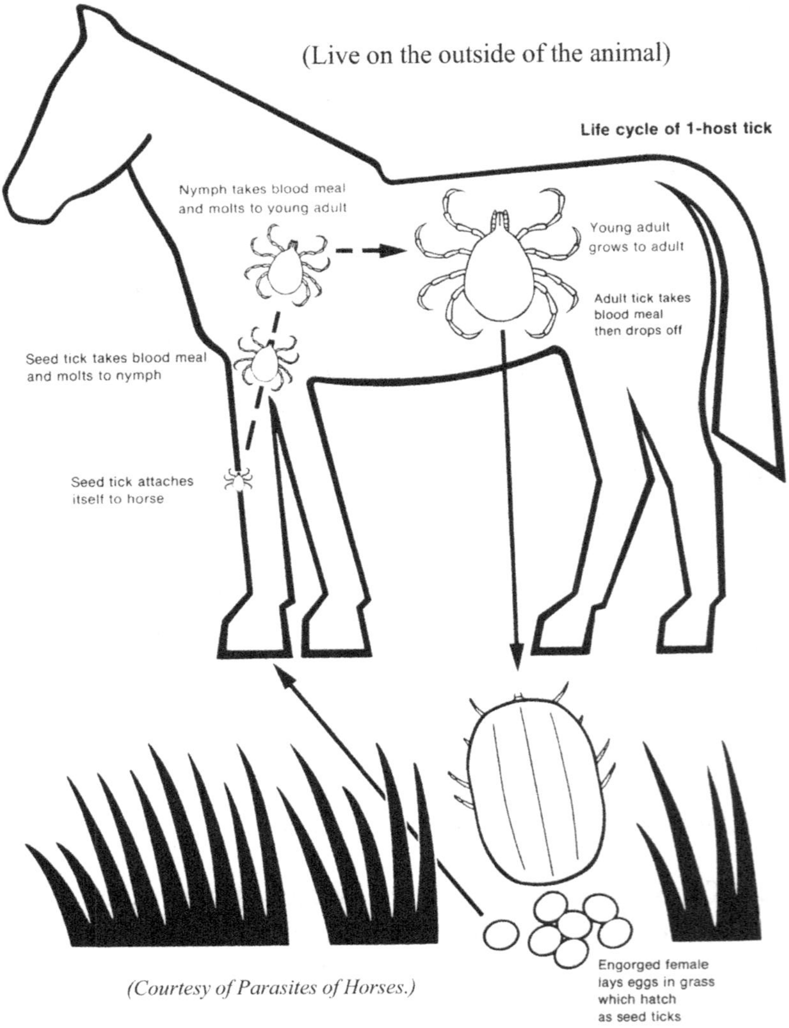

(Courtesy of Parasites of Horses.)

2. **Mites:** Mites can only be seen under the microscope. Mites eat skin debris, or puncture the skin and suck lymphfluid. Some forms burrow tunnels into the skin where they live, feed, and deposit eggs. Signs of mites (mange) include severe itching, hair loss, and scabs on the face or legs. Mange can only be diagnosed by scraping the skin and examining the debris under the microscope. Mange requires several treatments since new pests will hatch fromeggs that survive the first treatment.

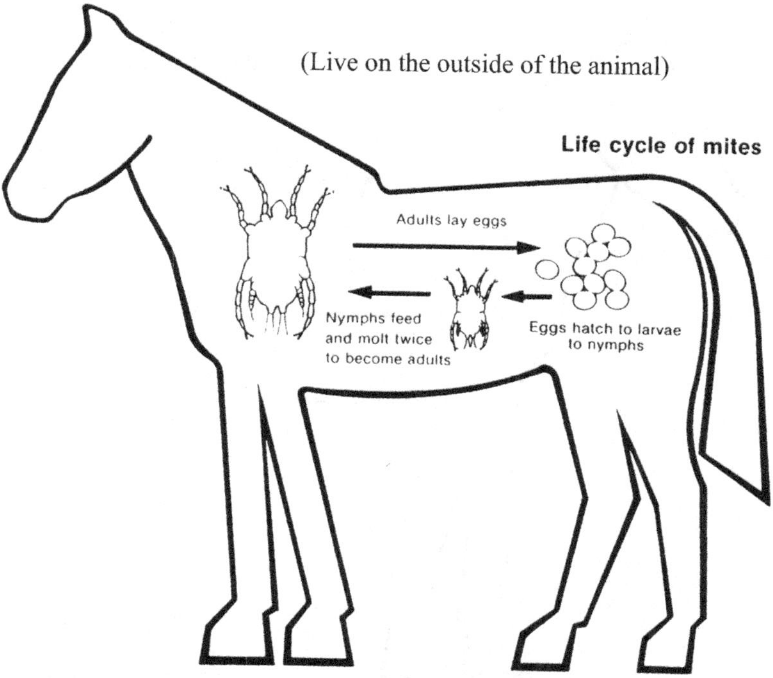

(Courtesy of Parasites of Horses.)

3. **Lice:** Lice are wingless, flattened insects with six legs adapted for clinging to hair. Lice occur in colder climates. Two kinds of lice are common. The sucking louse has mouth parts adapted for puncturing skin and sucking tissue fluids and blood. The biting louse feeds on skin debris and has mouth parts adapted for chewing. Lice are more common in the winter and spring months when the coat is the longest. Lice cause skin problems, blood loss, anemia and weight loss. The first sign of lice is itching and hair loss along the horse's mane, neck and tail. If youpart the hair and look carefully you will see little whitish moving dots. Lice are treated with spray or powder insecticide. Ivermectin will help kill the sucking lice as well. Treatments should be repeated in two weeks to kill newly-hatched larvae from the eggs present at the first treatment which were not killed.

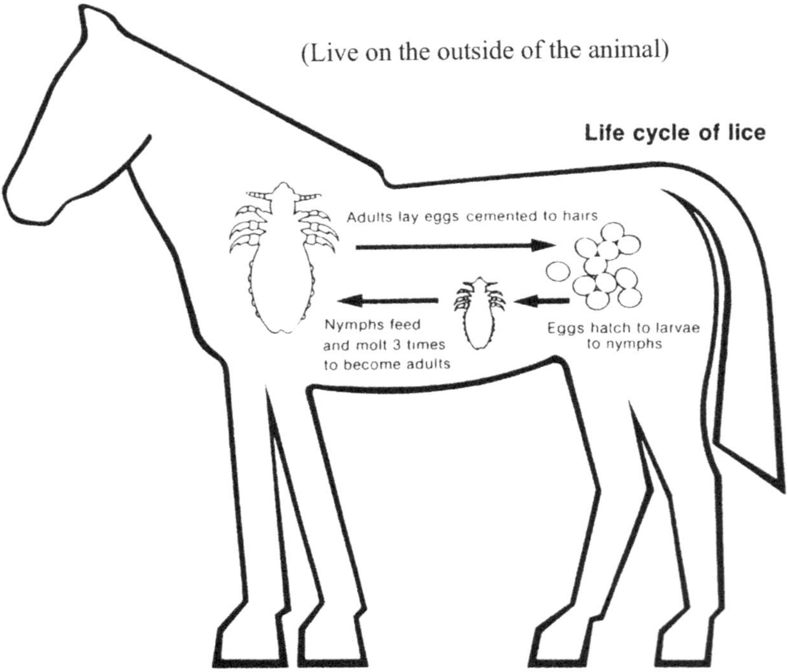

(Courtesy of Parasites of Horses.)

4. **Botflies:** Botflies lay their eggs on hairs of the head, neck or legs of horses. The horse becomes infected when the eggs are licked off the hairs. The larvae hatch and then develop in the tissues of the mouth before developing into an internal parasite which lives in the stomach over the winter. In the summer the bot larvae pass in the manure and develop into a fly which lays the eggs again on the horse.

5. **Horn Flies and Buffalo Flies**: These flies lay eggs in the manure. The eggs hatch into larvae that feed on the manure. They grow and molt into pupae that rest in the dried manure or in the ground. The pupae may overwinter or emerge as adult flies in a few days. The adults mate and lay eggs to commence the next generation. Flies can transmit disease and become so annoying that animals will not eat or drink.

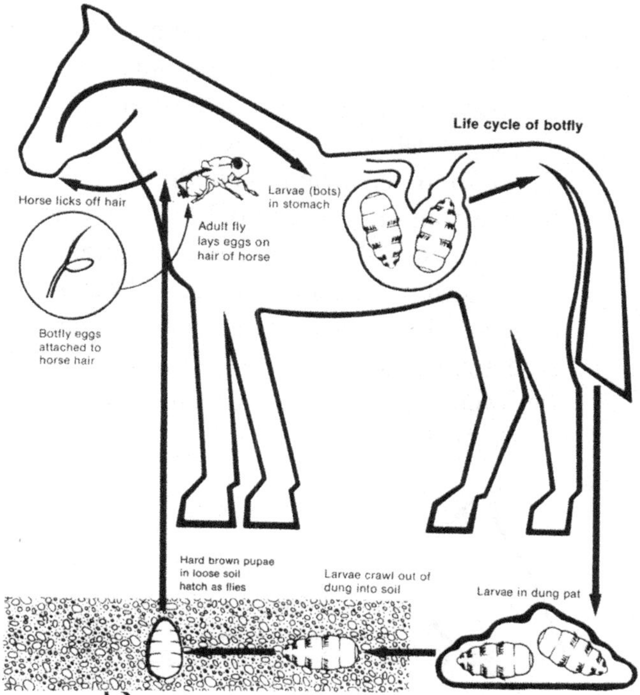

(Courtesy of Parasites of Horses.)

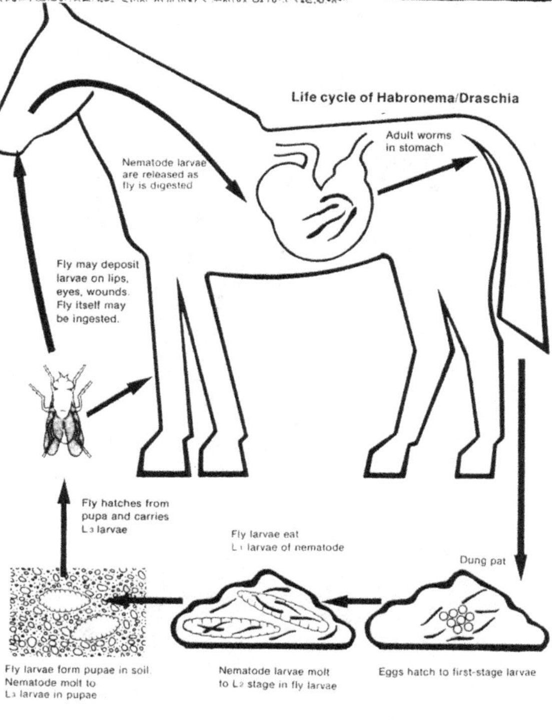

Internal parasites (nematodes) are often transmitted by an intermediate host (fly).

6. **Blow Flies and Screw Worms:** Blow flies are universal while screw worms are commonly seen in Central and Latin America. Adult flies lay eggs in any open sores on the horse. The eggs hatch into larvae that feed on the flesh of the host for 3 to 6 days and then drop to the ground where the transformation to the pupal stage occurs. Adults hatch from the pupae and the life cycle is repeated. Abscesses anddeath to the host can occur in a matter of a few days. Any wounds and surgical areas must be cleaned and treated with insecticides and screw worm paste to prevent problems. Petroleum jelly mixed with insecticide or use of Smear 62 works well.

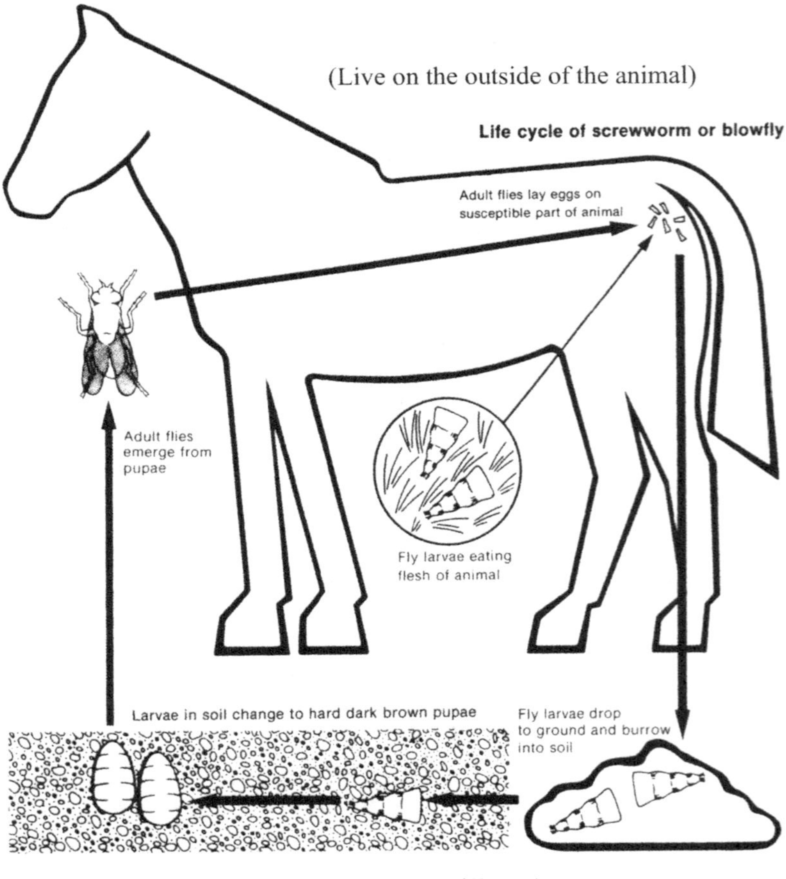

(Courtesy of Parasites of Horses.)

Ectoparasite Damage to the Host

1. **Skin Lesions**

- Parasites feeding on the host's skin may cause intense itching - especially with lice and mites.
- Mange is a skin disease caused by damage from mites. It is characterized by raw, thickened skin that has lost its hair. Much of the damage is actually inflicted as a result of the horse rubbing from the intense itching. Painful deep sores can result from the parasites and animal itching.
- The entire lesion is susceptible to secondary bacterial infections and attracts flies and screw worm damage.

2. **Loss of Blood**

- Sucking of blood by lice and ticks can be a serious drain on animals. Iron stores that are used to make red blood cells become depleted and anemia occurs.
- Animals may be forced to use ingested nutrients to replace losses rather than to build lean body mass or muscle.

3. **Transmission of Disease:** Equine viral encephalitis and piroplasmosis are transmitted by the tick. Parasites carry the infective disease organism from animal to animal between blood meals.
4. **Tick Paralysis:** Some ticks are capable of releasing certain toxins in the blood, causing a paralysis syndrome. The paralysis occurs with the hind limbs being affected first, followed by the forward regions. This problem is seen in North America, Australia, Africa, and India. The paralysis is more common in younger animals. Recovery is in a matter of hours if the ticks are removed in time. *Ixodes* and *Dermacenter* species are capable of causing this paralysis.
5. **Annoyance:** Animals plagues by external parasites become anxious and do not feed or rest well. The result of parasite worry may be loss of weight, condition, and stamina.

Ectoparasite Control

1. **Spraying and Dipping:** Many countries supervise the control of ticks and other ectoparasites with government programs of spraying or dipping. Contact the local government agency for help.
2. **Dusts and Powders:** These can be applied topically to help rid the host of external parasites.
3. **Pour-ons:** Organophosphates can be poured on the back of horses affected with mites. 2 ounces of Dicholrovous poured on the back of a horse will work as an effective treatment. The chemical resides in the tissues of the horse and kills the mites.
4. **Wound Care:** All wounds should be cleaned, disinfected, and treated with an insecticide to prevent ectoparasite infestation.
5. **Eye protection:** Masks or tassels attached to the halter can protect the eyes and prevent infection.

(Courtesy of Easy Health Care for Your Horse.)

6. **Ivermectin (Ivomec):** Some parasites that suck blood (like blood sucking lice and ticks) are susceptible to Ivermectin. Use the oral paste or give the injectable solution orally to the affected animals. Do not inject cattle Ivomec into horses as this can cause *Clostridial* infection and possibly death. Ivomec can be given orally although it is not as effective as the equine paste form.

7. **Mix insecticides with Petroleum Jelly:** Apply this on sores or around the eyes and the ears to protect from insects.

Insecticides are potentially dangerous drugs. Always follow directions closely, especially with mixing and dilution.

Treatments for ticks and lice may need to be repeated in10-15 day intervals to kill the new hatches of parasites.

Products for External Parasite Control

Co-Ral 25% wettable powder - Mix .5 kilo in 200 liters water for lice—pour or spray on the animal.
Co-Ral 1% Dust - Dust the animal for lice.
Lindane 12.4% - mix ¾ liter in 200 liters of water for mange, lice and some ticks. Pour or spray on the animal. Do not use on newborns.
Malathion 4.5% dust - good for lice and fair for mange. Dust the animal.
Malathion 57% emulsifiable concentrate - Mix 2 liters in 200 liters of water and spray or dip for lice and mange.
Sevin (carbaryl) dust - Very safe to dust for lice.
Rabon dust - Dust the affected animal.
Toxaphene 60% emulsifiable concentrate - Mix ¾ liter in 200 liters of water—effective for manage, ticks, and lice.Pour or spray on the animal.
Ivermectin (Ivomec) - Effective for some lice, mange, and blood sucking ticks; also kills the bot larvae in the horse. Do not inject Ivomec into a horse. It can be given orally if the paste is not available.
Neguvon (Trichlorfon) - pour on for external parasites.
Parmethrins - From the chrysanthomum flower. It repels flies. New formulations will also kill the flies.

Internal Parasites

Internal parasites or endoparasites live within a host and must obtain nutrients from that animal in order to survive and reproduce. The host's ability to thrive or even survive is often decreased as a result of the parasite presence. Endoparasites are divided into three categories: roundworms (nematodes), tapeworms (cestodes), and flukes (trematodes). Roundworms are by far the most important parasite of horses.

Nematodes - Roundworms

1. **Strongyles**: These are blood sucking worms known as blood worms or red worms. The worms are found in the intestines where they cause extensive damage to the blood vessels and mucous membranes. Colic from blood vessel damage can be deadly to the horse. Anemia from blood sucking is common in severe infections of the parasite. Young animals are more severely affected.

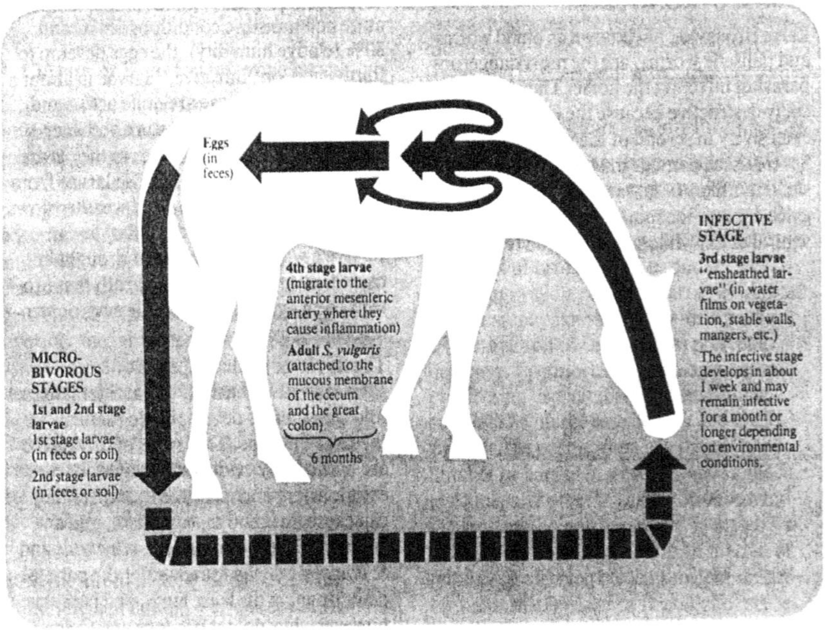

Life cycle of *Strongylus vulgaris*. The life cycles of S. *edentatus*, S. *equinus* and small strongyles are similar to the life cycle of S. *vulgaris* in the free-living stages but differ in the parasitic stages.

2. **Ascarids:** Commonly known as roundworms, they do most of their damage to foals and young horses. Adult horses develop an immunity to these parasites so they are not present in great numbers. Ascarids cause major lung and liver damage when the larvae migrate through thebody. The parasite can cause poor growth and development,respiratory infections, intestinal impactions and bowel rupture with ensuing death.

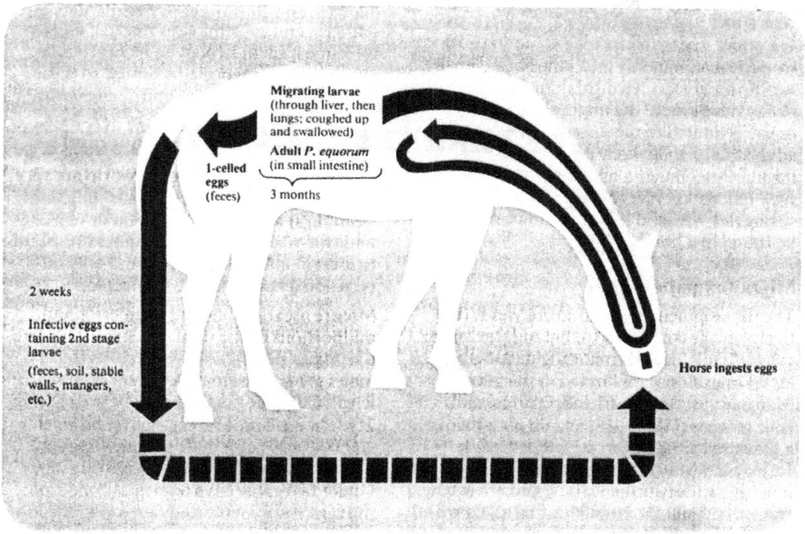

Life cycle of *Parascaris equorum (an ascarid)*

3. **Pinworms:** Pinworms live in the lower intestine. They lay their eggs around the anus causing the horse to itch and rub its tail. Pinworms are more of an annoyance than a major impact on the horse's health.

4. **Lungworm:** Donkeys are a common host and reservoir of infection for horses. If only a few numbers are present, little damage occurs but death can result in the face of large parasite infection. The parasite causes irritation of the bronchial mucousa (breathing tubes) leading to coughing, difficult breathing, and loss of appetite. Secondary pneumonia can occur. Dry pastures help prevent lungworm infection as the larvae are sensitive to drying.

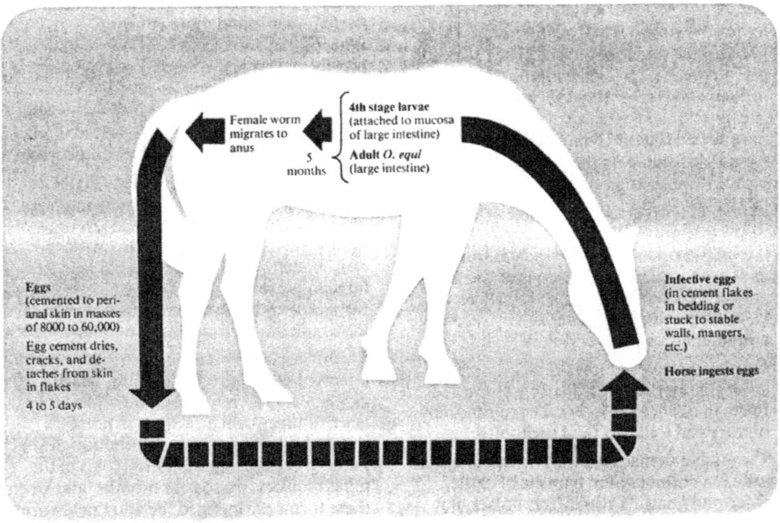

Life cycle of *Oxyuris equi (a pinworm)*

5. **Habronema:** This is a nematode that is carried by the housefly or stablefly. If the larvae are deposited in wounds or on the eye, they do not develop further but feed on the sores and cause more tissue damage. These woundsdo not heal and can cause disfigurement to the horse. Thebest control is wound care and daily treatment with insecticides plus periodic Ivermectin administration to kill the larvae.
6. **Onchocerca:** This parasite is transmitted by a *Culicoides* biting midge. Adult worms develop in tendons and ligaments of the horse causing swelling, pain, nodule formation and restriction of movement. Microfilaria can also infiltrate the eye causing irritation and blindness. This parasite is easily controlled by periodic use of Ivermectin.
7. **Thelazia:** This is a long slender worm that can reach 18 mm in length. The worm lives on the eye and in the tear ducts of the horse and is spread by the face fly. Heavy infections cause eye damage possibly leading to blindness.The adults can be removed under local anesthesia.Prevention is also dependent on sanitation, screening, and fly control with insecticides.

Cestodes - Tapeworms

Tapeworms are most significant if the host is already in poor condition. Large numbers can cause intestinal irritation producing diarrhea, colic and an unhealthy appearance. Large numbers can block the gut and be fatal.

(Courtesy of Parasites of Horses.)

Trematodes - Flukes

This parasite occurs in wet areas worldwide. The life cycle involves the snail as an intermediate host for development. The adult flukes lay eggs in the bile ducts which then pass in the feces and develop in the snail. After leaving the snail the immature forms encyst on grass and are eaten by the horse. The parasites then penetrate the gut and migrate to the liver.It is this migration of the fluke that causes damage to liver and associated tissues. Anemia can result along with decreased growth, weight loss and edema. Snails can be eliminated from pastures by draining the pasture or by applying copper sulfate.

Parasite Control

Management Practices

1. Cleanliness and Sanitation

- Keep stalls and pens dry and clean.
- Eliminate wet or muddy areas.
- Keep pens free from trash and debris.

2. Manure Disposal

- Pick up manure twice weekly and compost.
- Compost manure before spreading on pastures.
- Or spread on cropland and other land that will not grazed.

3. Pastures

- Rotate pastures or staked out areas as much as is practical.
- Graze young animals separately from older horses if possible.
- Alternate grazing horses with other species of livestock.
- Do not overgraze the pasture as this provides more contamination and exposure to parasites.

4. Feed and Water

- Avoid manure contamination in water and feeding areas.
- Keep feed up off the ground in bunks and feeders when possible.

5. Wound Care

- Clean and disinfect all wounds and use insecticides.

Treatment Programs:

1. Younger horses are more seriously affected by internal parasites than mature animals. Start treating foals when they are six weeks of age and treat them every six to eight weeks until they are two years old.
2. Mature horses need deworming every two to three months.
3. Alternate deworming products to avoid making parasites become resistant to the medicine.
4. Follow directions being sure to use the proper dosage.

Products to use:

Product	Trade Name	Use
• ivermectin	Eqvalan Zirmectrin	excellent for nematod & bots
• pyrantel pamoate	Strongid Imathal	cestodes use 2x dose, nematodes
• febendazole	Panacur	nematodes
• mebendazole	Telmin	nematodes
• cambendazole	Camvet	nematodes
• thiabendazole	Equizole	nematodes
• phenothiazine	Pheno-Sweet	poor against nematodes
• piperazine	Equivet Wonder Wormer	poor
• diclorvos	Equiguard	mites, nemtodes
• trichlorofon	Anthon, Combot	mites, nemtodes

- Ivermectins work well against most parasites except cestodes. They are also effective against some skin parasites and migrating larvae. If a horse is heavily parasitized, useone of the other products first, and then use the ivermectin2-4 weeks later to avoid a severe reaction from many dead parasites. **Do not inject Ivomec in horses.**

- The benzimidazoles (product names that end in "zole") products typically produce resistant parasites if used repeatedly. They do not easily kill migrating worms but are very safe in large doses.
- The pyrantel pamoates are safe and also useful to kill cestodes in three times the normal dose. This is a good product to rotate with ivermectins.
- Organophosphates can cause colic symptoms so the dose must be carefully followed. This product should be avoided in mares that are more than 5 months pregnant. The liquid form of the product can be poured on the back (2oz to an adult animal) to treat mange. This product is also effective against bots.
- Thiabendazole can be used for resistant nematodes if ivermectin is not available. Dose the animal at 10 times the normal dose (440mg/kg) daily for two successive days.
- Phenothiazine and piperazine are not highly effective parasite products. It would be better to use the above products as a first choice if they are available.
- Deworm all new animals before they are introduced to the herd.

Note—This parasite control program follows conventional strategies and conceptions. New control programs emphasize fecal egg counts to identify high shedders. See *Addendum A* in the back of the book for more information. Realistically, in developing countries, any worming program is better than none. Also, labs may not be easily accessible.

Section 9

Feet & Shoes

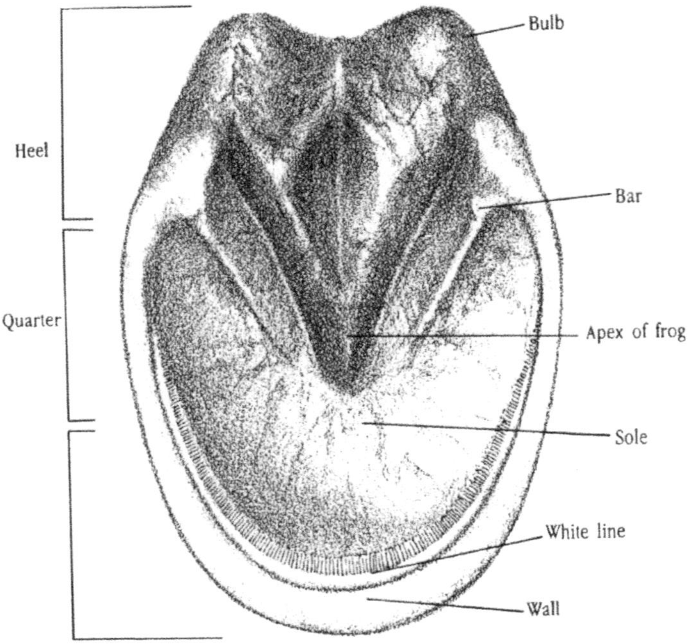

Someone once said, "no hoof, no horse." Sound feet and proper care are very important in the health and function of horses, mules, and donkeys. Constant observation and examination of the feet along with proper care is most important.

Wild horses do not have caring owners and farriers to keep their feet in perfect condition. However, wild horses do not encounter the same conditions as domestic horses. Hooves of the wild horse wear down at a rate roughly equal to the speed of growth and its diet contains plenty of fiber and minerals to keep the horn strong. Domestic horses on the other hand, lead a restricted life. Turned out, it is bounded by the limits of the field or tether, often on soft soil that does not wear the feet. In work, its exercise is likely to be partly on roads. Long hooves can develop, then crack and split without proper care.

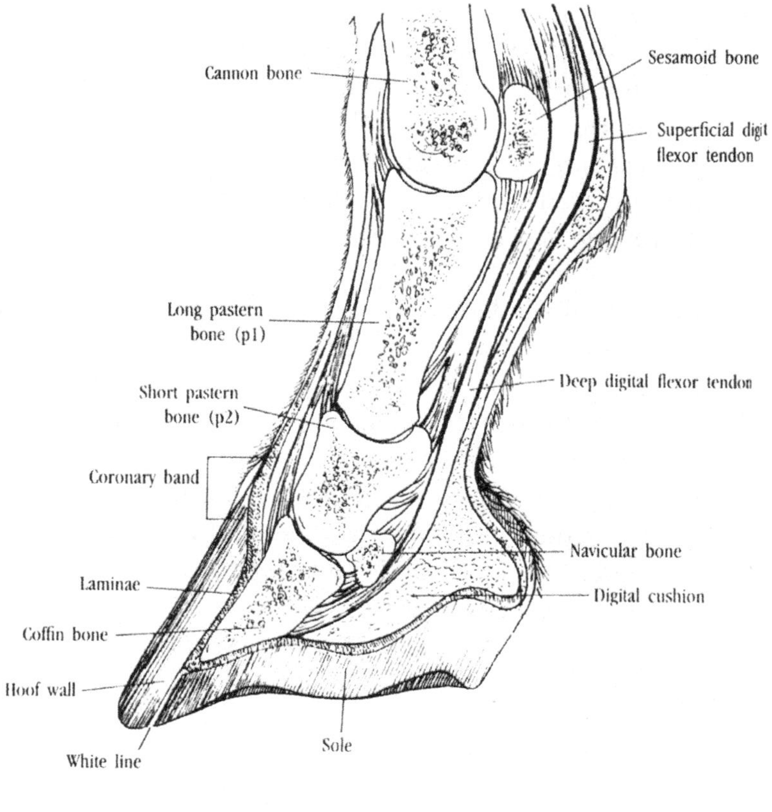

Hoof Structure

1. **The Wall:** This is the part of the hoof visible when the foot is on the ground. It grows downwards from the coronet just like a finger-nail. The wall embraces the foot, and at the heels is inclined inwards to form the bars. The outer surface of the wall has a glossy, varnish-like finish that prevents undue evaporation from the horn. This outer surface should never be rasped off! This prevents the horn from becoming hard and brittle. The toe, quarters, and heelall form part of the wall of the foot.
2. **The Sole:** This protects the foot from injury from underneath. It is rather thin for this purpose, and liberties cannot be taken with it. In its healthy state the sole is slightly concave, like a saucer turned upside down which helps to provide a better foothold.
3. **Frog:** This is nature's anti-slipping and anti-concussion (shock-absorber) device. The frog is the first part of the foot to make contact with the ground and plays an importantrole in ensuring a good foothold. Its wedge shape, irregular surface and central cleft help its anti-slipping function. The frog's effectiveness as a shock-absorber stems from its size, its rubber-like consistency, its upward flexibility; and the cushion within the foot upon which it rests.
4. **Interior of the Foot:** The interior of the foot is made up of bones, joints, and sensitive structures any or all of whichare liable to injury should the wall, sole, or frog be penetrated.
5. **Shape & Position:** The ideal shape and position of the hoof is shown below.
6. **Pastern:** The slope of the pastern, if carried through the hoof to its junction with the ground, should form an angle of 45-50 degrees to the ground. This slope should beparallel with the shoulder. Remember, horses are not perfect in conformation and severe deviations can lead to lameness with work.

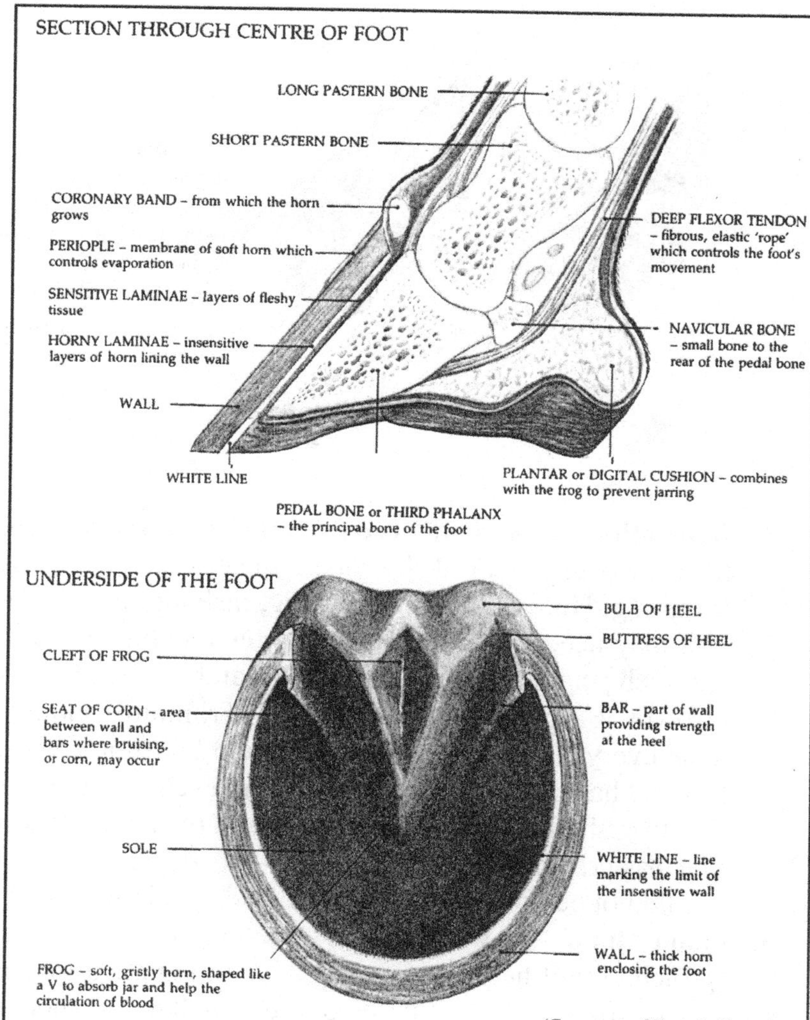

(Courtesy of Feet & Shoes.)

Foot Care

1. **Pick the Feet:** The feet should be picked out daily. A bent screw driver at a 90 degree angle works well for this procedure. Pick out any rocks and all debris. Look for punctures, nails or rocks puncturing the sole. Examine the frog looking for black, smelly material which would indicate thrush (p.128).

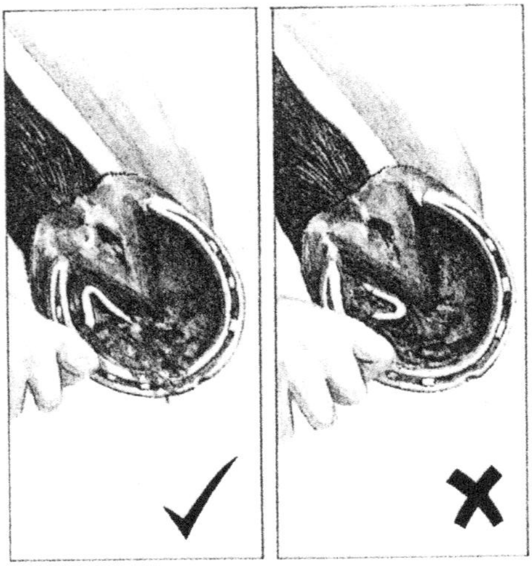

Use a **hoof-pick** to clean debris from the sole and around the frog. Always work the hoof-pick away from you for fear of damaging sensitive tissue. *(Courtesy of Feet & Shoes.)*

2. **Determine if the hoof needs trimming:** Long toes, high heels or overgrowth of the shoe indicates the hooves need trimming. Hooves that are cracking, dishing out or curling also may need trimming. Hooves grow one-quarter to one-half inch per month depending on diet and season of the year (growth is slower in cold seasons). Trimming should be done every six weeks. The foot should be trimmed so the foot is "balanced." A balanced foot is even on both sides. The frog should contact the ground with the horse standing. The pastern angles should be 45-50 degrees. The toes should not be left too long, nor the heels too high or low.
3. **Good hoof growth:** Feeds containing amino acids (proteins) will help hoof quality. Lanolin applied topically is good for hooves. Oil-based products will dry the hoof out and promote cracking.
4. **Determine if the horse need shoes:** Horses on soft ground without cracking feet and who are not put under stress of work do fine without shoes, but they still need regular trimming. With work, especially on hard surfaces, shoes may be needed. Shoes may also be indicated when certain hoof abnormalities or leg problems are present. When shoes are used, they should be reset every six to eight weeks.

Shoeing

Remember the adage, "no foot, no horse." Improper shoeing can be more harmful than going without shoes. Use someone experienced in proper hoof trimming if they are available. Shoeing protects the wall from wear plus it protects the sole from bruising and damage when horses are worked on hard or rocky ground. The following diagram illustrates the basic farriertools needed in shoeing a horse:

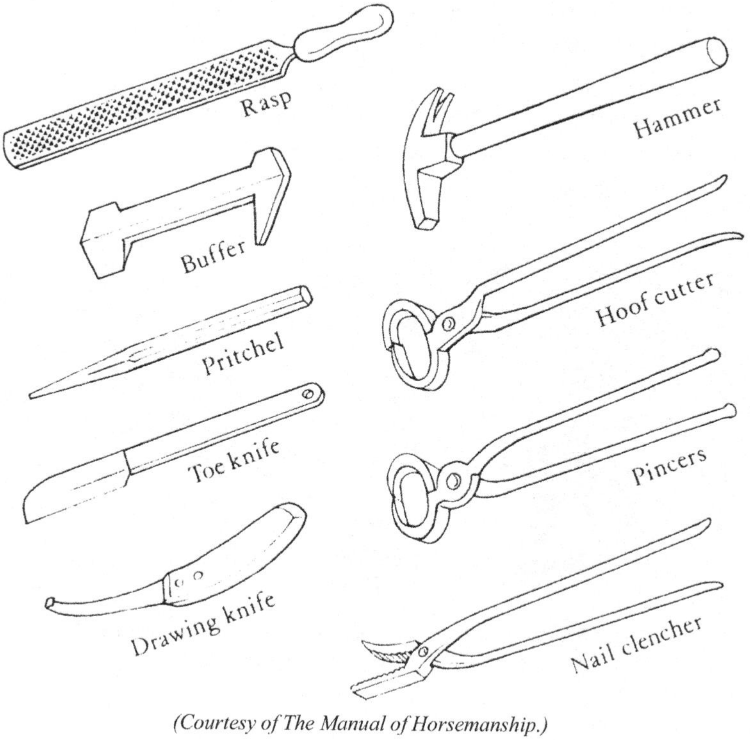

(Courtesy of The Manual of Horsemanship.)

Re-shoeing

1. **Refit or reshoeing** is done when the hoof has grown sufficiently to require resetting of the shoe. With this procedure the shoe is removed, the foot trimmed and the shoe is reset on the foot. This will prevent toe length from becoming excessive causing the horse to stumble.

2. Indications the horse needs to be reshod or refitted

- Clenches (horseshoe nail pieces) have risen and stand out from the hoof wall
- The foot is over-long and out of shape
- Some part of the shoe has worn thin
- The shoe is loose or has come off
- The hoof wall has grown over the shoe

3. Removal of the shoe

- First cut all the clenches using a buffer and driving hammer or rasp them off
- Then lever the shoe off with pincers.
- If the clenches have been cleanly cut or rasped, there should be no breaking or tearing away of the wall as the shoe is released from the hoof.

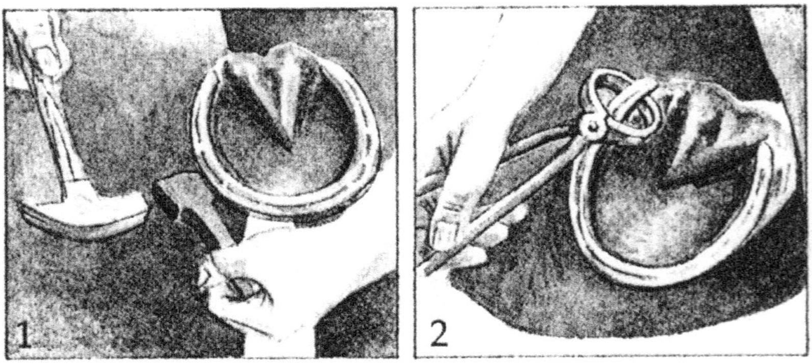

1. Cutting off the clenches with **hammer** and **buffer.**
2. Using the **pincers** to lever off the old shoe.

(Courtesy of Feet & Shoes.)

4. Preparation of the Foot

- Reduce the over-growth of the wall using a toeing knife or hoof cutter and prepare the foot for fitting of the previous shoe or a new shoe.
- Trim away any of the old sole and ragged parts of the frog.
- Use a rasp to give the foot a level bearing surface.
- Make sure the foot is balanced and even in all respects

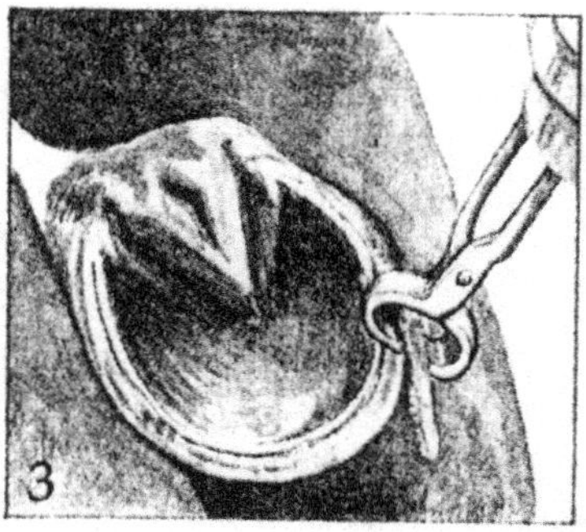
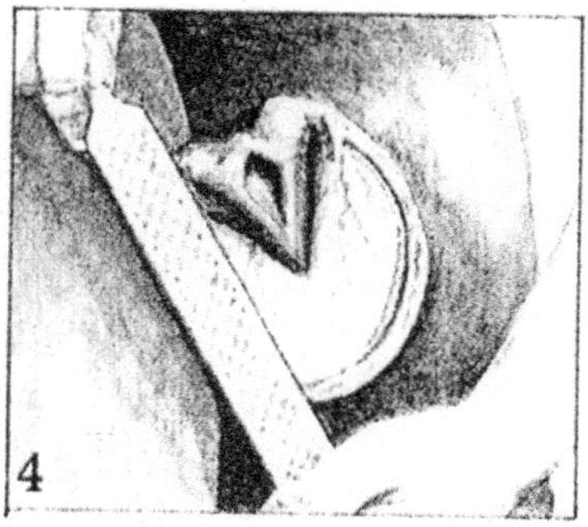

3. The **hoof cutter** trims excess growth of horn to reveal the white line.
4. The **rasp** levels the surface of the foot to receive the shoe.
(Courtesy of Feet & Shoes.)

5. Forgeing

- In making a shoe, the weight and type of iron selected depends on the work the horse will be doing. After shaping the iron to the foot (do not shape the foot to the shoe), nail holes are stamped and clips are made.
- With manufactured shoes, pick the correct size to fit the foot or adjust the shoe so the fit is correct.

6. Fitting

- The shoe must fit the foot. Improper fitting can cause lameness.
- If an attempt is made to fit the foot to the shoe, often contraction of the heel, bruising, or loose shoes and lameness may occur.

7. Nailing On

- The shoe is cooled (if heat has been used) in water and nailed on.
- The first nail is usually driven in the toe. Nails are made in various sizes and it is important to use the correct size. Nail heads too large will wear away too soon and those too small will not fill the hole…both of which will lead to a loose shoe.
- The ends of the nail where it penetrates the wall should be turned over and twisted off. This leaves a small piece called a "clench."

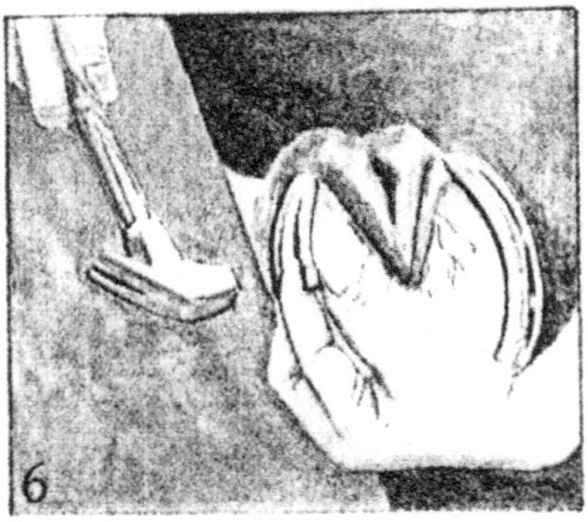

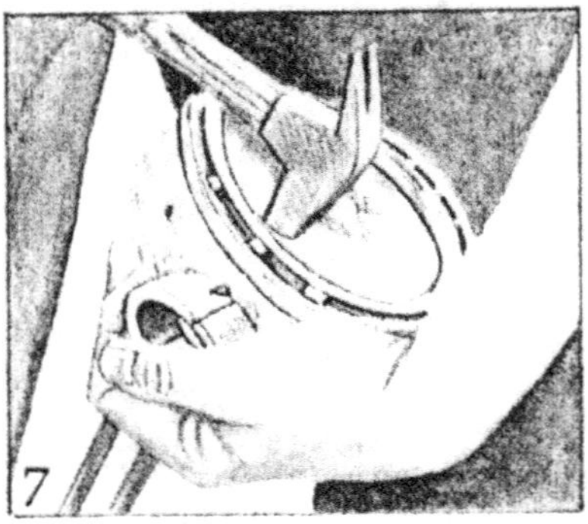

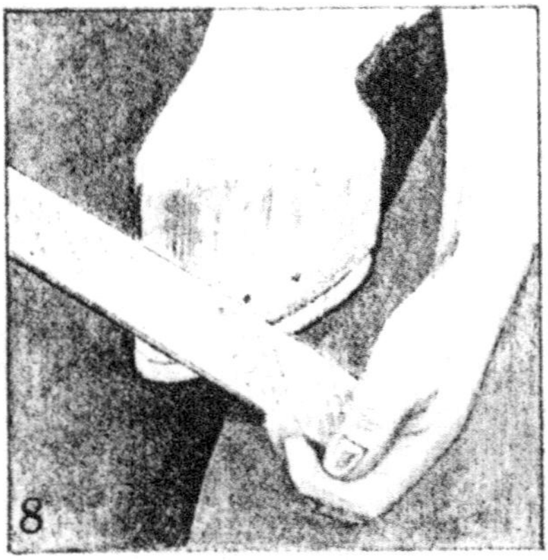

6. The **nails** must be angled correctly.
7. The emerging **nail tip** is twisted off and bent down to form the **clench.**
8. The foot is finished off by **rasping** below the line of clenches.

(Courtesy of Feet & Shoes.)

8. Finishing

- The clenches are tidied up with a rasp and a small indentation or bed is made for them in the hoof wall with a hammer.
- To reduce the risk of cracking, a rasp is run around the outer edge of the wall where the shoe and wall meet.

Common Faults

1. The foot was made to fit the shoe: The wall should not be rasped away excessively to meet the iron. The toe should not be dumped (over rasped).
2. The type of shoe is unsuitable for the work of the animal.
3. The foot has not been reduced in length at both the toe and heel or on each side so that it is not level or balanced.

4. The surface of the frog is not in contact with the ground.
5. The sole has not been trimmed to a concave shape but bears weight on the ground.
6. The walls have been trimmed too short and the horse is sore.
7. The nails are not the correct size.
8. The nails have been inserted too high and into the sensitive laminae so the horse becomes lame with an abscess.
9. There is daylight between the shoe and the foot, especially in the heel region.
10. The bars of the shoe extend onto the frog or sole causing lameness or corns.

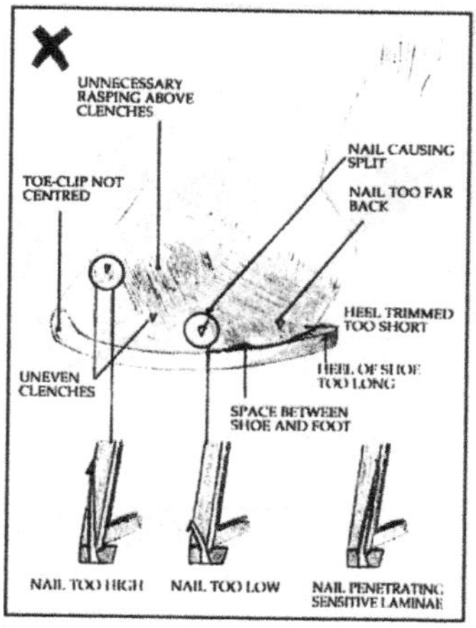

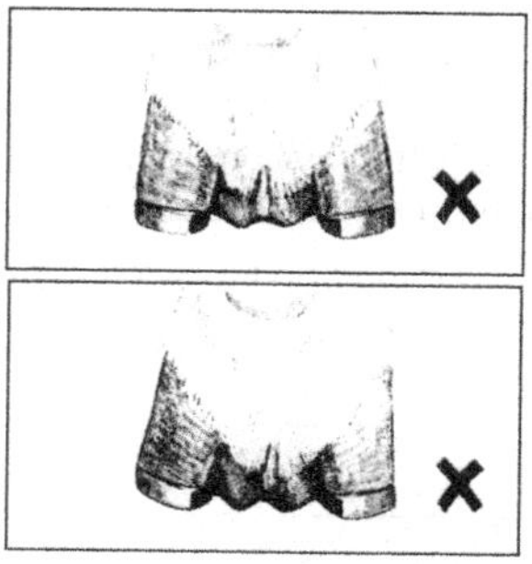

Poorly made shoes and **badly prepared feet.** *Top:* the frog does not touch the ground. *Bottom:* the shoe is uneven and the foot is crooked and not square. *(Courtesy of Feet & Shoes.)*

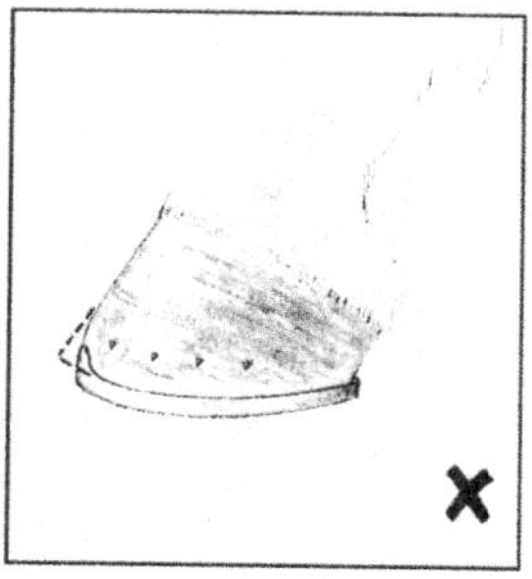

Dumping, in which the foot is shortened to match an ill-fitting shoe by rasping the toe, is a serious fault and can result in brittle feet. *(Courtesy of Feet & Shoes)*

Types of Shoes

1. **Plain-stamped:** This is the simplest form of shoe. It consists of an unmodified bar of iron shaped and stamped with nail holes. This shoe is suitable for horses doing relatively slow work since it has no provision against slipping.

2. **Studded shoes:** These shoes help provide traction and prevent slipping. Studs are usually fitted on the outside of the hind shoes.

- Studs worn on the front shoes are apt to put strain on the tendons so this is not advised. Road studs are permanently fitted into the shoe. They are made of slower-wearing metal and do not protrude too much from the shoe so the balance of the foot is hardly affected.
- Screw-in studs must be removed as soon as the horse has worked and the hole filled with cotton wool to prevent them from becoming clogged.

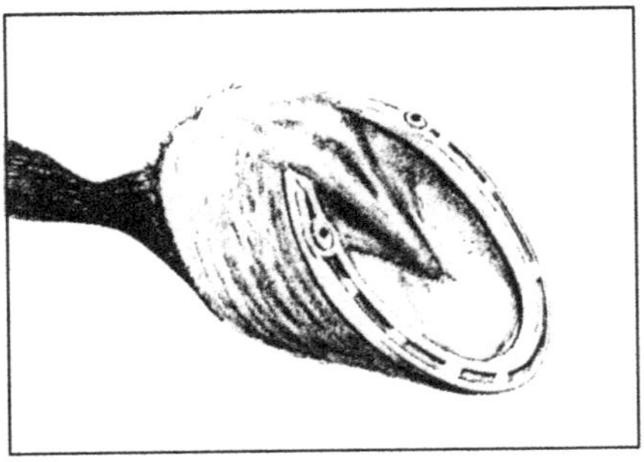

To upset the balance of the foot as little as possible, two studs or nails are better than one, and they should protrude by only a small amount.
(Courtesy of Feet & Shoes.)

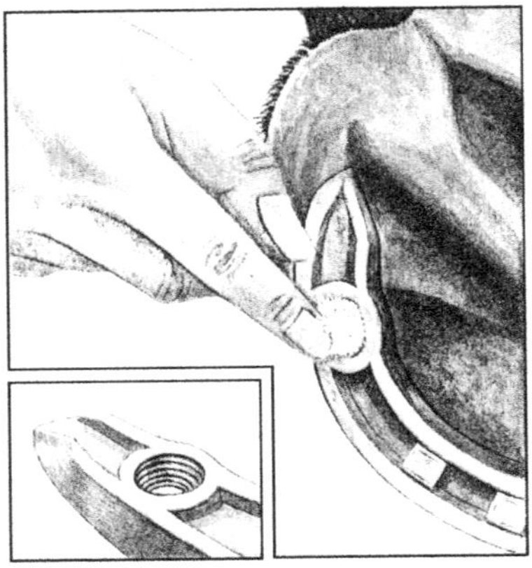

Screw-in studs fit into a threaded hole stamped into the heel of the hind shoe. The hole should be packed with tow or cotton wool when not in use.
(Courtesy of Feet & Shoes.)

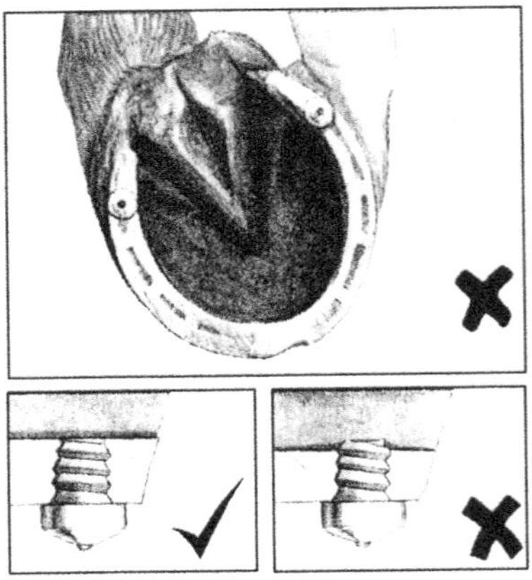

Top: Take care not to use screw-in studs on a **worn shoe,** for fear of damaging the foot. As shoes wear down *(bottom),* so, of course, does the depth of the hole for the stud.
(Courtesy of Feet & Shoes.)

3. **Padded shoes:** With some injuries such as bruised soles or punctures that have been treated and are healing, a leather or neoprene rubber pad is indicated under the shoe to protect the foot. The feet must be kept dry when pads are used and the pad should not be left any longer than neces- sary to avoid problems with the frog and sole. Pine tar and okum, a similar fibrous substance, or silicone applied under the pad will help add further cushion and prevent moisture from collecting, which could lead to thrush or canker.

There are many other types of shoes but in developing countries these may not be available.

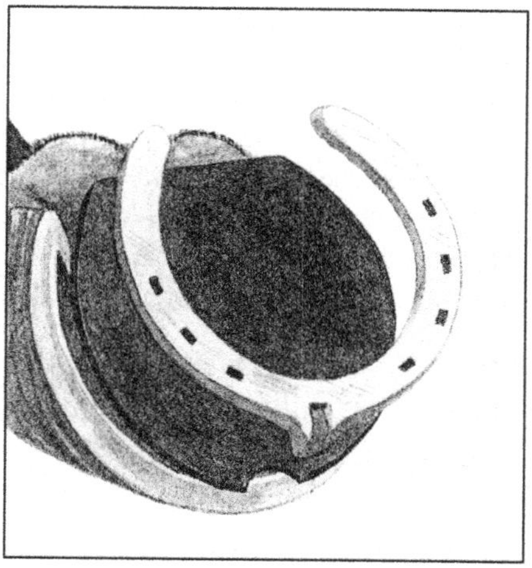

A **leather pad,** shaped to the foot, may be placed between shoe and foot to protect bruised or punctured soles where infection has developed.
(Courtesy of Feet & Shoes.)

The Problem Foot

Cracked Feet

1. Signs

- Dry flaky hooves
- Cracks at the toe or quarters
- Soft hooves

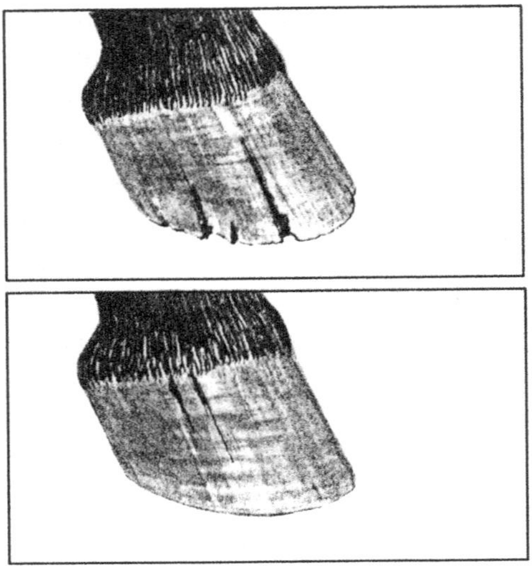

There are two types of crack. The **grasscrack** *(top)* may develop in over-long feet. The **sandcrack** *(bottom)* is more serious and harder to treat.
(Courtesy of Feet & Shoes.)

2. Treatment

- Trim the feet on a regular basis. Long toes tend to cause the foot to flare and break off or crack.
- Improve the protein nutrition.
- Apply non-petroleum based dressings with lanolin, fish oils, vegetable oil, and pine tar to a wet hoof each day.
- Avoid rasping the hoof wall and tearing away its protective covering.
- If the horse is constantly in mud, move them to a well drained area. If the climate is very dry, let the water trough over flow a little to give the foot exposure to moisture.
- Use a clip on a shoe (see picture below) to help remove stress on the cracked area. Rasp the cracked area so it is not touching the shoe.
- Deep cracks may need to be drilled out, sewn together, then filled with fiberglass or supported with small sheet metal screws and a strip of metal laid across the crack.

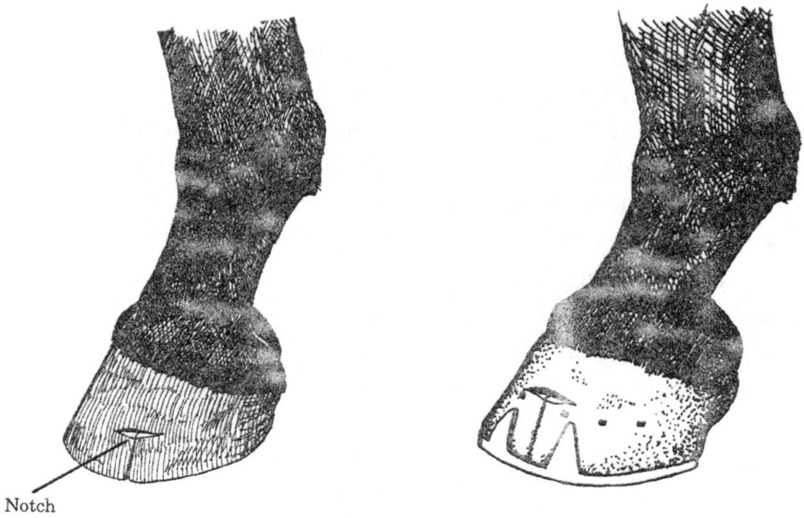

Treatment of toe crack originating at bearing surface. The hoof wall is grooved or notched to prevent further cracking and the edge of the crack is trimmed short to prevent pressure from the shoe. Toe clips are used on either side of the crack to support the wall and prevent expansion of the crack under pressure.

View of bar shoe with clips applied to immobilize a bearing surface heel crack.

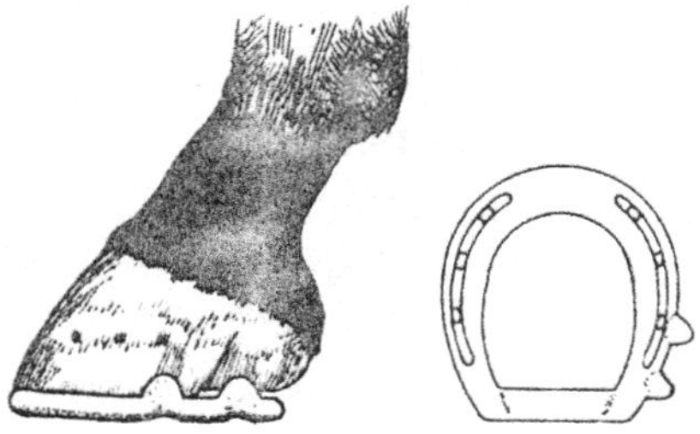

(Courtesy of The Illustrated Veterinary Encyclopedia.)

Thrush or Canker - These are infections of the foot and are a common problem especially in wet climates or pastures.

1. Signs

- Thrush is a foul smelling black material in the grooves of the foot.
- Canker is an infection of the sole and frog and extends into deeper tissues.
- Both cause varying degrees of lameness

2. **Treatment**

- Keep the animals in dry, well drained areas.
- Clean all dead, infected tissue away.
- For thrush, daily applications of liquid bleach, iodine, or formaldehyde will kill the bacteria.
- For canker, Penicillin may be indicated for 5 to 10 days.
- Remove dead tissue and pack with iodine or Betadine.

3. **Prevention**

- Keep animals in clean dry areas.
- Clean the feet daily.
- Treat any thrush early with the above treatment.

Foot Bruise - Trauma from riding on rocks or hard surfaces can cause bruises to occur. These bruises can progress to abscesses in some cases.

1. **Signs**

- The horse is limping or sore, especially on hard surfaces.
- A reddish mark on the sole may be noticeable when the area is washed clean.
- The horse flinches when pressure is exerted on the area with tongs (hoof testers) or tapped with a small hammer.

2. **Treatment**

- Clean the feet, removing any rocks.
- Avoid hard or rocky surfaces; rest for a week or so.
- Apply shoes with a pad (refer to padded shoes and cautions).
- Some bruises are caused by improper fitting shoes when the bars of the shoe extend over the sole or frog. In these cases, the shoe needs to be reset properly.

Gravel or Sole Abscess – These occur when a rock or foreign body punctures the sole of the hoof, or a piece of gravel works its way up the white line. An infection is the result of this problem.

1. **Signs**

- Pain when putting the foot down
- Pain with hoof tester on the bottom sole in the area of the abscess

- Swelling above the hoof
- Pus discharge from above the coronary band
- Hole in the bottom of the foot

2. **Treatment**

- Treatment is aimed at draining the area with a large hole made at the site of entry in the sole.
- Soak in Epsom Salts water twice daily for a week.
- Bandage the foot and keep the foot clean and dry between soakings (see illustration on page 131).
- Tetanus Toxoid booster
- Penicillin injections for 1 to 2 weeks

3. **Prevention**

- Keep nails out of the barnyard.
- Put shoes on animals if they are working on sharp rocky ground.

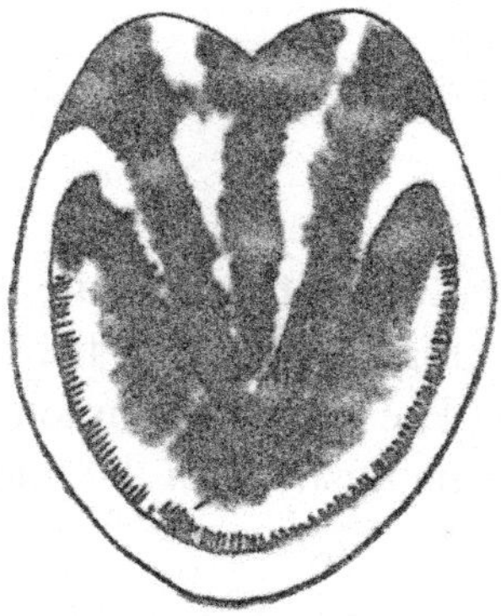

Illustration shows penetration of white line and travel of infection up to the coronary band. Drainage and enlargement of the wound promotes healing.
(Courtesy of The Illustrated Veterinary Encyclopedia.)

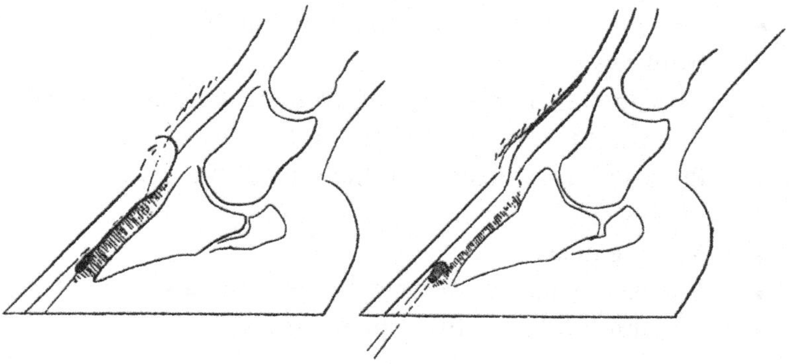

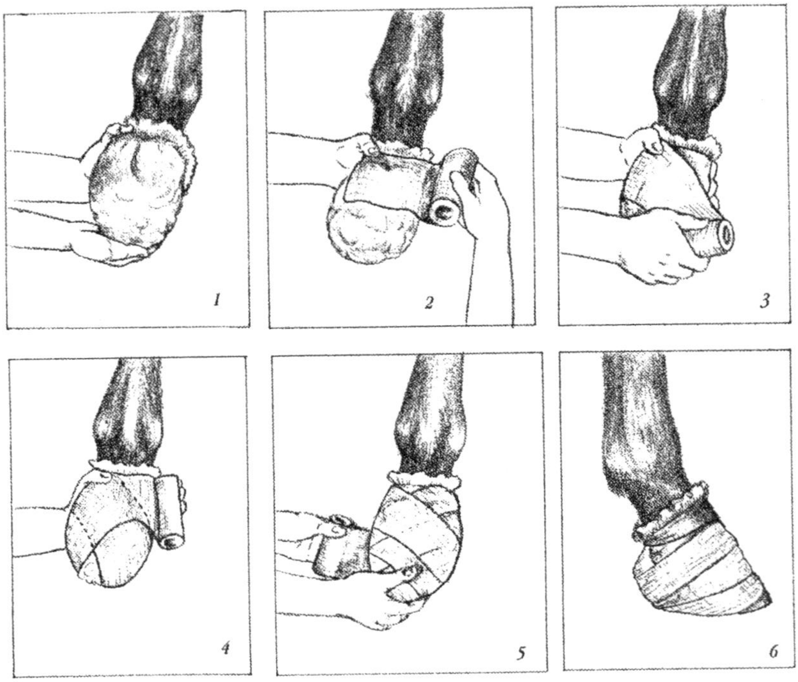

(Courtesy of Easy Health Care For Your Horse.)

Laminitis (Founder) – This is an inflammation of the lamina that lie between the hoof wall and the coffin bone of the foot. The swelling sometimes becomes so great that the bone can be pushed out the bottom of the foot by the intense pressure.

1. Signs

- Extreme pain, reluctance to walk, excessive lying down, difficulty in rising
- The horse is reluctant to bear weigh on the toes of the front feet.
- Classic stance is when the front feet extend forward and the weight is adjusted to the back legs.
- Pain with hoof testers on the toes of the feet.

2. Causes

- Many causes
- Excessive hard work or galloping on hard ground
- Getting too fat on lush spring grass

- Engorgement of grains
- Retained placenta (fetal membranes)
- Toxins

3. **Treatment**

- Eliminate the cause.
- Stand on soft sandy ground.
- Phenylbutazone twice daily
- Acepromazine in small injections 4-6 times daily to reduce blood pressure to the foot
- Rasp the toe off and put a horse shoe on backwards making sure that there is no toe pressure at the ends of the shoe.

(Courtesy of Easy Health Care For Your Horse

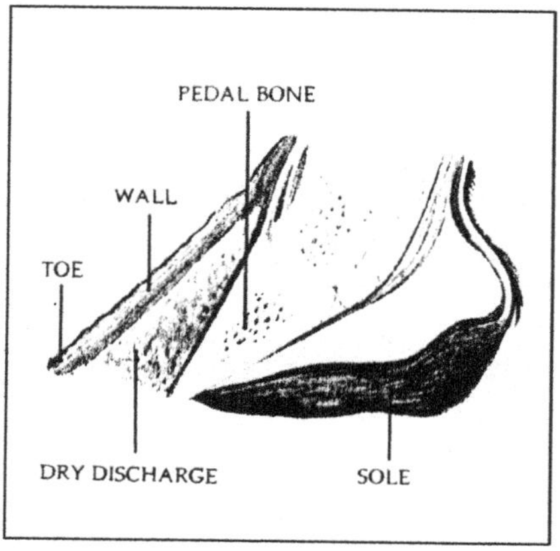

Following a severe attack of laminitis, **discharge** from inflamed laminae pushes the toe outwards. The swivelled pedal bone pierces the sole, which has dropped.

(Courtesy of Feet & Shoes.)

Section 10

Lameness

Detection, diagnosis, and treatment of lameness is often a very difficult process. Diagnosis of some problems require equipment and procedures beyond the scope of this booklet and resources available in primitive cultures. Often time and rest are the best treatments in some cases. Again, prevention of problems is very important. Selection of stock with good conformation (see Section 5) may help prevent many problems.

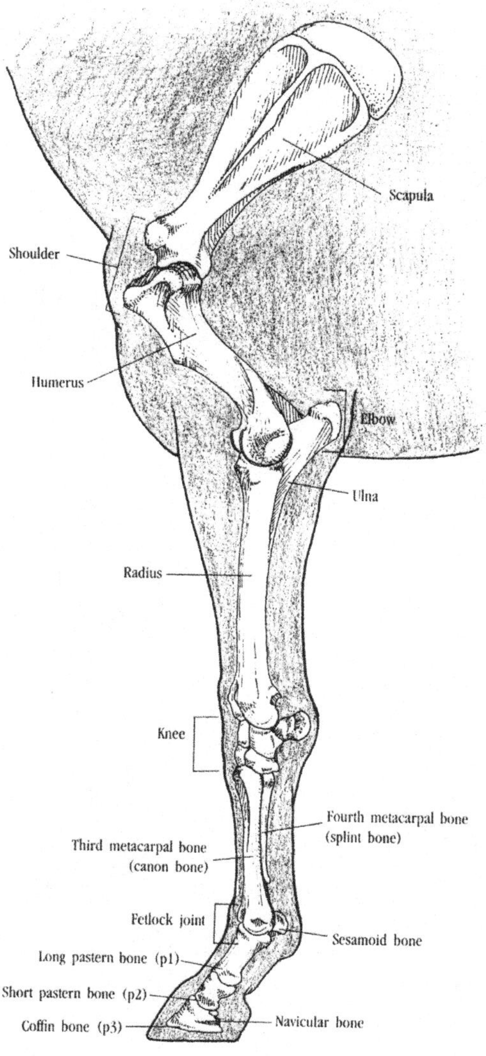

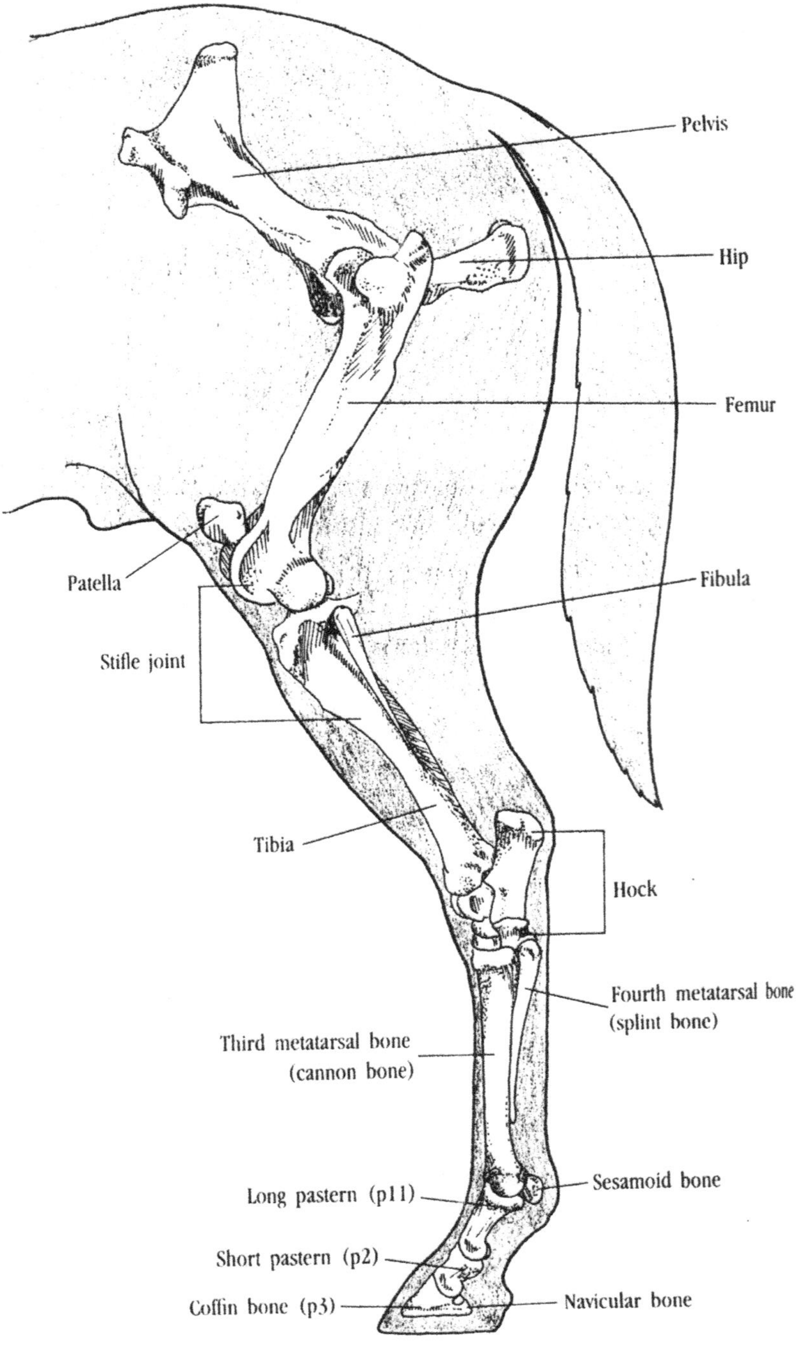
Pelvis
Hip
Femur
Patella
Fibula
Stifle joint
Tibia
Hock
Fourth metatarsal bone
(splint bone)
Third metatarsal bone
(cannon bone)
Long pastern (p11)
Sesamoid bone
Short pastern (p2)
Coffin bone (p3)
Navicular bone

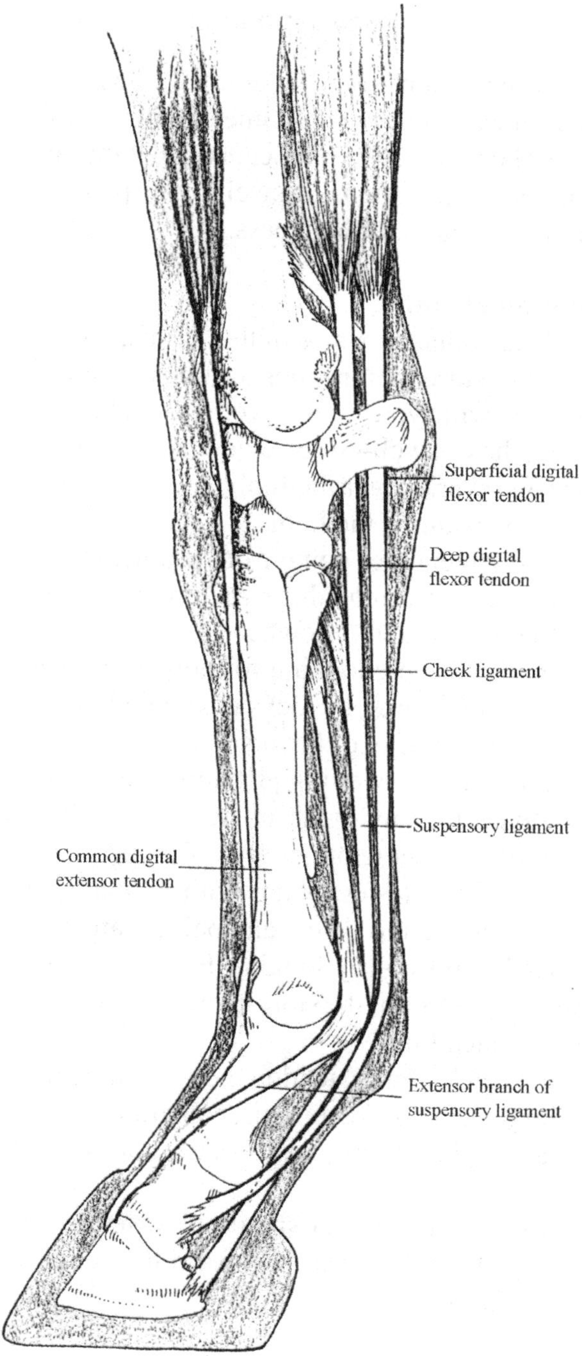
Superficial digital flexor tendon
Deep digital flexor tendon
Check ligament
Suspensory ligament
Common digital extensor tendon
Extensor branch of suspensory ligament

Examination

History: Information from the animal's owner can often give us clues as to why the equine is lame or having difficulty moving. Duration of lameness, when it occurs, what the animal is doing when sore, and indications of swelling or punctures all help in determining the cause of lameness.

Examination at rest:

1. Visually examine the limbs of the animal, noting any swelling, unusual posture, deformities, or unusual silhouettes.
2. Palpation with your hand over the limbs is the next step. Feel for heat, swelling, or evidence of painful response as your hand works over the limb.
3. Physically examine the hoof.
   - Most lamenesses come from problems in the hoof so be sure to examine this structure even when you see an obvious problem elsewhere.
   - Inspect the sole looking for punctures or foreign bodies like nails, sticks, or rocks embedded in the hoof.
   - Note any cracks or breaks in the hoof wall.
   - Look for any swelling just above the hoof.
   - Palpate for any areas of heat using the back of your hand to detect temperature differences.
   - Use a hoof tester or tap with a small hammer on the sole and wall to detect any painful areas.
4. Flex each joint tightly looking for reflex pain responses or decreased range of flexion in that joint. Compare similar joints in each limb.
5. Palpate each flexor tendon below the knee or hock for swelling, pain, or heat. This palpation should be done in both a weight-bearing stance and in a non weight-bearing stance.
6. The elbow and shoulder should be extended, flexed, pulled out laterally and pushed in toward the center of the animal looking for pain.

7. Use an ink pen and press down the back and over the rump on the midline and each side of the midline for pain response. Some animals will flinch due to apprehension, not pain.

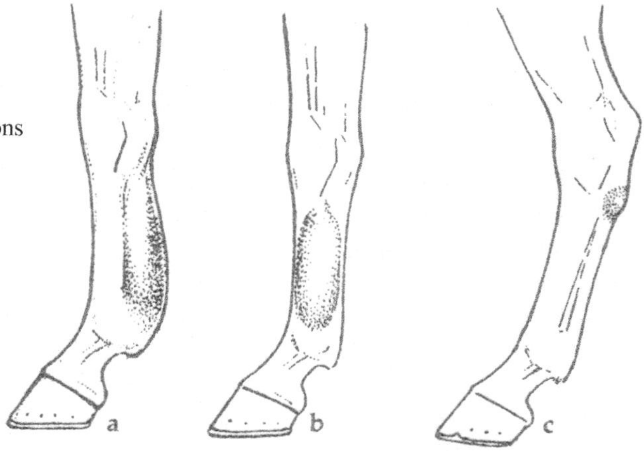

Sprains
(a) Sprained tendons
(b) Sprained suspensory ligament
(c) Curb.

Examination at Exercise: A lame horse is reluctant to bear weight on the sore limb. Perform lameness exams on level ground. Use both soft and hard surfaces if possible, avoiding rocky areas. Examine the horse from the side, from behind going away and from the front as it walks toward you. Also have the horse trotted in a small circle of four meters going each direction several times.

1. If it is a front limb soreness, the animal will bob the head down when the good leg hits the ground and bob the head up when the sore limb hits the ground.
2. When examining a lame horse trotting in a circle, a sore front limb will show the most head bobbing when the limb is on the inside of the circle. With rear limb soreness, the lameness will show most when the hind limb is on the outside of the circle.
3. With hind limb lameness view the animal from the rear. The animal will drop the hip when the good leg bears weight and "hike" or raise the hip and hock when the sore leg bears weight.

4. You will often notice a shorter stride or extension of the foot on the lame leg when viewed from the side.
5. Observe the horse going and coming. Notice how each foot lands. Is it landing square or more to one side of the foot? Watch the hip and head for clues to soreness location.
6. Flex each joint for 90 to 120 seconds then trot the horse twenty feet in a straight line. If the horse shows more lameness after flexion of a specific joint, this joint may be a problem.
7. Walking a horse down an incline can demonstrate patella (see picture p. 135) problems of the hind limb if the limb extends straight back.

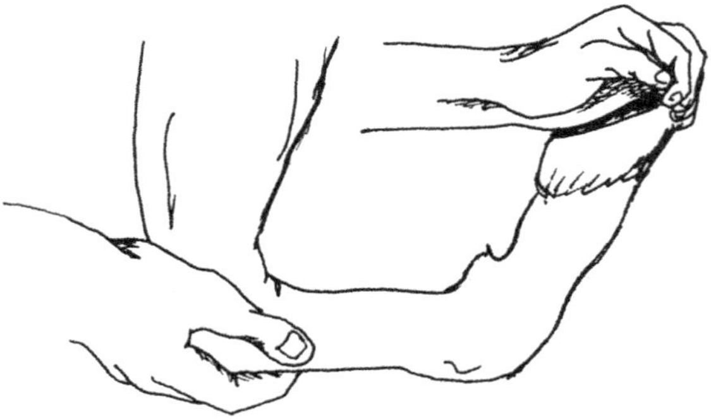

The interphalangeal joints (pastern and coffin) and the fetlock joints are being flexed. A painful response can implicate any one of the joint spaces.

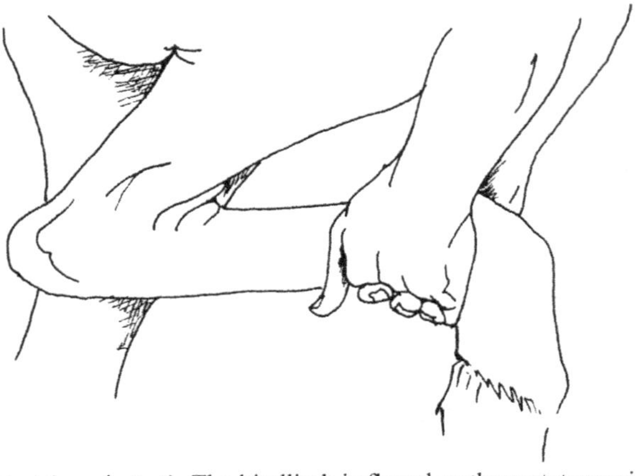

Hock flexion test (spavin test). The hindlimb is flexed so the metatarsus is approximately parallel to the ground surface. It is held in this position for 1 to $1^{1}/_{2}$ minutes, and the horse is observed for increased lameness. Increased lameness is considered a positive test but not specific only for spavin since the stifle and fetlock joints are also flexed.

Diagnostic Aids: It is sometimes difficult to find the source of lameness even after a thorough physical exam, and further steps are needed to help with a diagnosis.

1. Local Nerve Blocks

- Nerve blocks will deaden an area below the spot where the local anesthetic is given. If the animal is lame after the nerve block (and the animal shows no pain sensation to needle pricks in the area) then the lameness is above the point of the nerve block.
- 1 to 5 cc of lidocaine is injected over a nerve with a 25 or 22 gauge needle. A time span of 6 minutes is given for the anesthetic to work. Trained individuals are often needed for success of this method.
- Nerve blocks are very helpful in localizing lower leg soreness and differentiating them from shoulder problems of the front limb.

2. Radiographs (x-rays)

- Although they may not be available in many lands, x-rays can be helpful in determining causes of lameness, especially after a nerve block has determined the specific area of soreness.
- X-rays are used mainly to distinguish hard tissue problems such as those with bones. They can give clues to some soft tissue problems like joint linings, or ligament problems where they attach to bones.

Joint Fluid Examination

- Joint fluid can be taken from a swollen joint. Extreme care must be used to avoid contamination or infection of a joint when doing this procedure. The area must be clipped, scrubbed, and disinfected with betadine and alcohol. Sterile syringes and needles must be used.
- If the joint fluid is filled with pus and cloudy, you can suspect an infected joint.

Response to Medication and Rest

- Especially in isolated areas, the best diagnostic technique available may be simple rest. Rest coupled with poultices to draw down swelling, or anti-inflammatory products like phenylbutazone can be helpful in some situations (if available and affordable) .
- Absence of response to rest may indicate a poor prognosis.

Diseases of Exertion

Exercise Related Muscle Damage (Exertional Myopathy): Also called "tying-up" disease or "Monday morning disease," this is an inflammation of the muscles of work in the back and hind limbs.

1. **Signs**
   - shortening of gait, stiffness, or reluctance to move
   - sweating, shaking, obvious discomfort
   - sore on palpation of the back and/or muscle spasms
   - mildly elevated body temperature, 104 F or 39 C.
   - coffee colored urine
   - shortened gait and muscle atrophy with chronic cases
2. **Treatment**
   - immediate rest, **do not move or work the animal** until signs clear
   - muscle massage, use DMSO and aloe vera if available
   - phenylbutazone 1-2 mg per kg twice daily
   - withhold grain and rich feed
   - selenium and vitamin E supplements may help
   - MSM (DMSO metabolite if available) given daily, orally
3. **Prevention**
   - Do not feed grain when the animal is not being worked.
   - Avoid sudden strenuous exercise.
   - Slowly work the animals with increasing loads and time of work to give them a conditioning effect.

Heat Exhaustion: Working animals in high humidity and high temperatures can result in heat exhaustion.
(See also Section 12, First Aid).

1. **Signs**
   - high body temperatures above 105 F. or 39 C.
   - bright red membranes
   - elevated pulse and respiration
   - profuse sweating

2. **Treatment**
   - Cool the animal in a stream or pour water on the head and lower extremities. Do not pour water on the back and rump.
   - Rest
   - Keep out of sun. Build protective cover if shade is not readily available.
   - phyenlbutazone 1-2 mg per kg orally twice daily
   - no feed or water until it cools down to normal
3. **Prevention**
   - Condition properly. Avoid excess body fat.
   - Avoid work with high humidity and high ambient temperature (situations of temperature and humidity combined above 140). Work during the cool hours of the day and give adequate rest.
   - Give plenty of water during work (in small frequent quantities).
   - Provide access to good salt licks.

Exhaustion after Work

1. **Signs**
   - often elevated temperature
   - relaxed anal sphincter
   - unusual rapid respiration movements that coincide with the heart beat may occur
   - very listless, lethargic, reluctance to move
   - may be reluctant to eat
   - dehydration (neck skin is slow to lay flat after pinching)
2. **Treatment**
   - Rest
   - These animals may be suffering from lack of electrolytes (especially potassium) or depletion of their energy stores. Provide a good salt; add "Lite Salt" (if available for potassium), provide a good feed.
   - In serious cases intravenous (IV) electrolytes and fluids may be needed to save the animal.

3. Prevention
- Mix "lite salt" (potassium chloride) with regular salt in a 1:2 ratio.
- Provide adequate feed and rests throughout the work.
- Condition the animals prior to heavy work.
- Feed animals to avoid thin body weights.

Muscle Strain: With hard pulling, carrying a heavy or unbalanced load, muscle strains can occur. This is an inflammation and/or tearing of the muscle fiber.

1. Signs
- Temperature is usually normal.
- Animal is lame, short-strided, or walks stilted.
- The affected muscle is sore on palpation - warm and hard.

2. Treatment
- Rest until pain subsides.
- Massage.
- Topical application of DMSO with aloe vera is often helpful.
- Give 1-2 gms per kg of phenylbutazone orally.
- Once healing is occurring, start physical therapy on the muscle by walking slowly, without any loads, gradually increasing the distance.

3. Prevention
- Condition all work animals before subjecting to heavy work.
- Allow rest periods during work.
- Feed a good diet and salt-mineral mixture.
- Balance all loads and avoid excess loads.

Sweeney: This is the atrophy (shrinkage) of two muscles in the shoulder. Other names are slipped shoulder, shoulder atrophy, or suprascapular nerve paralysis. Blows to the shoulder or pressure from a harness collar can cause nerve damage resulting in temporary paralysis or complete severing of the nerve.

1. **Signs**
   - difficulty extending the shoulder, a swinging gait
   - flattened shoulder area
   - prominent scapula spine (p. 134) with hollowed areas on each side of the spine
2. **Treatment**
   - Long-term rest
   - Determine the cause and prevent further nerve damage. For example, if a draft horse, be sure the harness collar fits properly.
   - Phenylbutazone, massage and liniments may help minor cases where there is no significant nerve damage.
   - Poor prognosis for cases with significant nerve damage as the nerve does not regenerate.
3. **Prevention**
   - Have all harness collars fit properly on the neck without undue pressure back on the scapular region

Sprains

Sprains of tendons, ligaments, and muscles around the joints often cause lameness. They may occur under any condition. Severe sprains may put a horse out of work for many months. Sprains may be preceded by minor swelling or heat from a minor injury but become major with more work, so daily vigilance is very important.

Sprained Tendons: Tendons are tough, fibrous, slightly elastic cords that attach muscle to bone. When the muscle contracts, the tendon moves the bone. Tendons may alsosupport joints. Sprains are a tearing of the tendon fibers

1. **Signs**
   - heat, swelling in the affected area
   - soreness on pressure by palpation
   - lameness of the affected limb
   - reluctance to put the heel down in the case of strained flexor tendons

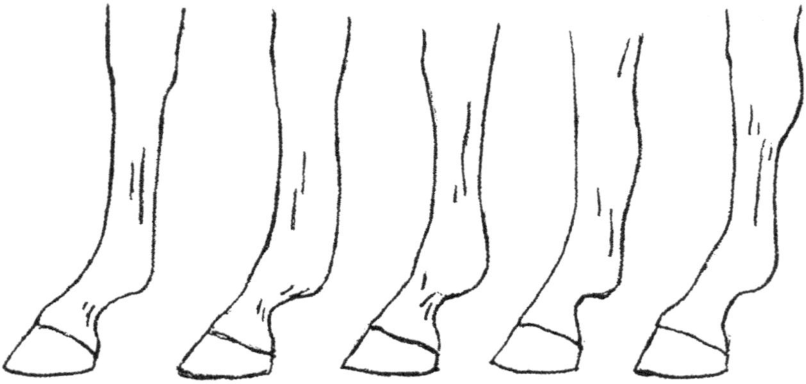

normal (healthy) | classic | low | medium | high

Sprained Tendon Examples

2. **Treatment:**
   - rest, up to a year or more in severe cases
   - standing in a cold stream, or cold soaks in acute cases
   - Support the limb using bandages.
   - DMSO topical or under a bandage
   - Once healing is thought complete, start exercise by slow walking on flat terrain with gradual increase in distance.

Sprained Ligaments: Ligaments are strong, fibrous bands which connect bones to bones and give support to joints. With the exception of the suspensory ligament, they do not stretch. The same signs and treatment for sprained tendons also applies to sprained ligaments.

Bursal Enlargements: Bursae are fluid sacs containing synovial fluid that lubricates joints or tendons so they move without friction. Bursae also protect bony points such as the elbow or point of the hock. A strain or trauma will cause excess fluid to be produced and soft swellings will result. Strains are caused by the sudden or sustained pulling on a joint or tendon.

Types of Bursal Strains

Windgalls or windpuffs: These are fluid swellings just above and to the sides of the fetlock joint between the flexor tendons and the suspensory ligament. They are not usually associated with pain but form with work in some individuals.
Bog Spavin: These are soft swellings on the inside of and to the front of the hock joint.
Thoroughpins: These are soft swellings in front of and slightly above the point of the hock between the Achillestendon and the bone.

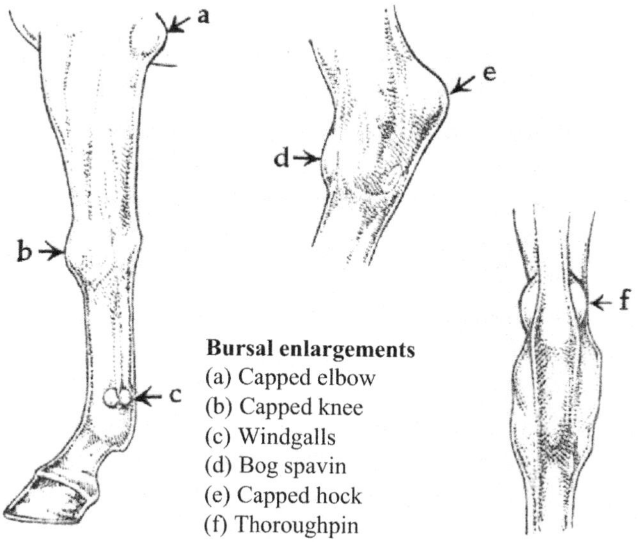

Bursal enlargements
(a) Capped elbow
(b) Capped knee
(c) Windgalls
(d) Bog spavin
(e) Capped hock
(f) Thoroughpin
(Courtesy of Easy Health Care)

Treatment of Bursal Strains

- Standing in a cold stream to reduce the swelling and inflammation may be helpful.
- Rest if there is any soreness.
- Massage with DMSO may help.
- Bandage limb for support if sore.

Some bursal swellings do not resolve, especially with windgalls. If there is not any soreness and the swelling does not enlarge, treatment may not be needed.

Types of Bursal Injuries

Joint Swelling: If the fetlock, carpus, hock, knee, elbow or shoulder has swelling involving all the area around the joint, that joint has a problem. This type of swelling may or may not be associated with pain. The problem may be caused froma bone chip, inflamed joint tissues (synovitis), inflamed or damaged joint ligaments, or by infection.

1. Examination

- Some infections of the tissues just below the skin will look like a swollen joint; be sure to differentiate between the two problems.
- Look for any punctures that might indicate an infection.
- Little or no lameness indicates more of a mild inflammation. An infection or severe structural damage may be associated with severe lameness.
- A joint tap, taken by sterile procedure, will help differentiate between an infected joint and an inflamed joint. Do not tap the joint if the tissue surrounding the joint may be infected.

2. Treatment

a) Infected Joint

- This joint should be flushed with sterile saline using a sterile procedure, with sterile syringe-needle & gloves.
- Give penicillin (40 cc/500 kg twice daily) and gentamiacin (4.4 mg/kg once daily) for two to four weeks.
- This treatment may be beyond the scope of expertise, economics, and product availability. In this case the animal should be destroyed.

b) Strained Joint

- Rest while sore.
- Cold soaks or stand in a cold stream.
- Apply DMSO topically.
- Apply support wraps.
- Give 1-2 mg/kg of phenylbutazone twice daily.

b) Bone Chips or Severely Damaged Ligaments

- Long term rest
- Same short term treatment as strained joints
- Animals unable to work may be evaluated for use in breeding.

Capped Elbows: These are soft swellings under the skin on the point of the elbow, usually caused by lack of bedding or from rubbing with the heel of the shoe on the elbow when lying down.

Capped Hocks: These are soft swellings at the point of the hock usually caused by persistent kicking against a solid object like a wall.

1. Treatment

- Rest if soreness exists.
- Cold soak the first few days then warm soak thereafter.
- Prevent further trauma by providing bedding, padding the area, and preventing further kicking.
- Topical DMSO may help reduce the swelling.
- Give 1-2 gm/kg of phenylbutazone to reduce swelling.
- Some swellings may not come back to normal even with extended treatment.
- If the skin is broken, refer to the wound notes in Section 12 for proper care.

Bony Enlargements

Splints: These are small bony knobs which form on the splint bone and/or the cannon bone below the knee or hock. Splints can be inflamed tissue (the ligament between the splint and cannon bone) or a crack in the splint bone. Splints are caused by direct trauma or concussion (hard work on hard surfaces). Poor conformation (bench knees) or improper shoeing (heel or hoof wall too high) predisposes horses tosplints.

1. **Signs**
    - Heat, pain on palpation, hard swelling
    - Lameness varies according to the degree of inflammation. Some never show soreness but fractures are usually painful.
    - Swelling shrinks in size as inflammation abates.
    - Lameness may persist if the exostosis (new bone growth) presses on tendons or the carpal (knee) joint.

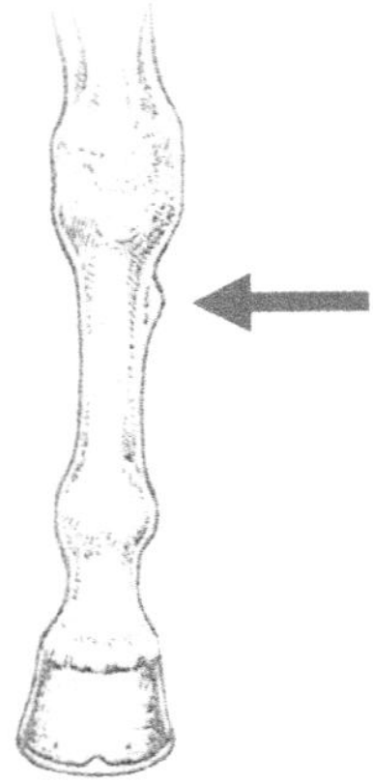

Splint

2. **Treatment**
    - Rest as long as swelling is present.
    - Cold hydrotherapy (cold stream or soaks).
    - Topical DMSO may help reduce inflammation.
    - If the splint impinges on the joint or a tendon, treatment may not help.
    - If the bone is fractured, long periods of rest may be needed. Surgery, the best treatment, may not be available.
3. **Prevention**
    - Proper balanced hoof trimming and shoeing
    - Let animals mature before hard work.
    - Avoid individuals with poor conformation .

Bone Spavins: These are bony enlargements on the lower part of the inside of the hock caused by strain, concussion, or poor conformation. There is a degeneration of parts of the joint and resultant arthritis with eventual fusion of the joint. The condition is often caused by faulty conformation (cow or sickle hocked, or narrow thin hock conformation). Sprains, strains or trauma to the hock can also play a part in the formation of bone spavins.

1. **Signs**
   - enlargement on the inner side of the hock
   - variable degrees of lameness with increasing lameness as the arthritis and exostosis progress
   - heat is seldom found in the problem area
   - decreased joint flexion, positive flexion (spavin) test

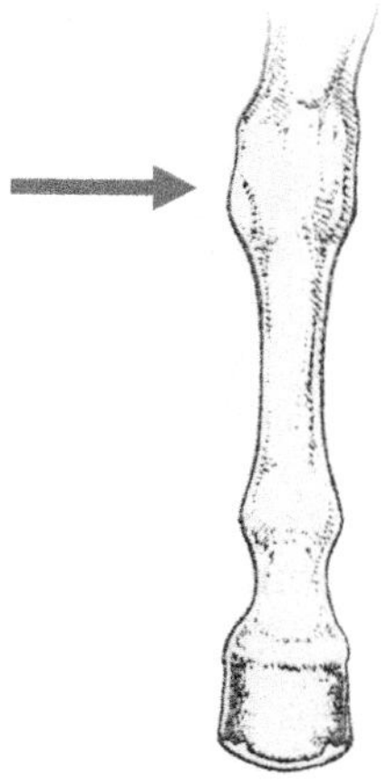

Bone spavin (hind leg).

2. **Treatment**
   - rest
   - moderate work until the joint fuses; then the horse should be sound again (may take years)
   - Roll the toes when timing or shoeing.
   - Sterile intraarticular injections of steroids may help the lameness.

Ringbone: Ringbone is the bony enlargement with new bone around the pastern or fetlock joint. The end result is fusion of the joint over a long period of time. Trauma, constant concussion on hard surfaces, and poor conformation (basewide, base narrow, toe in, toe out) can lead to ringbone.

1. **Signs**
   - slow, hard enlargement around a joint
   - lameness developing gradually
   - poor flexion of the affected joint
   - positive flexion test

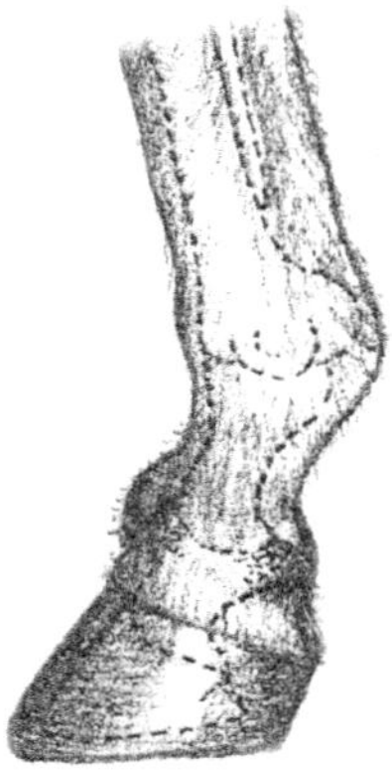

Ringbone

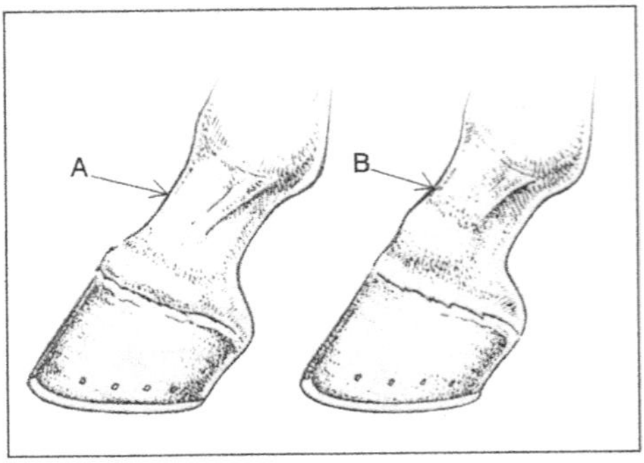

(a) Low ringbone. **(b)** High ringbone

2. **Treatment**
   - See treatments for bone spavin (previous page).
   - If only the pastern joint is involved, once joint fusion occurs, the horse can be used again. With other joints, the mechanical stiffness makes this difficult.

Bone Fractures & Joint Dislocations

Due to the economics in developing countries, most fractures and dislocations cannot be treated. If the bone is merely cracked, rest and bandage support with sufficient diet and time may allow the animal to heal. If the fracture is unstable, the animal should be destroyed. Bone chips in the joints can result injoint swelling and eventual degenerative joint problems. Surgery again is expensive and may not give good results.

Bone chips that are not in a joint but on a bony surface can cause a chronic draining wound. These can be treated by scraping the dead bone away and administering flushes and antibiotics. Again, this care may be available and/or expensive.

Other Conditions

Upward Fixation of the Patella: This condition, called stifled, occurs when the stifle joint (knee) is stuck while fully extended. The patella becomes fixed on the end of the femur by the patellar ligaments. The cause of the problem can be bone cysts, poor joint cartilage, poor conformation, or dam- aged ligaments.

1. Signs

- Stifle and hock are locked in a fully extended position.
- The leg cannot be flexed and the toe drags on the ground.
- Fixation may come and go, or be almost constant.

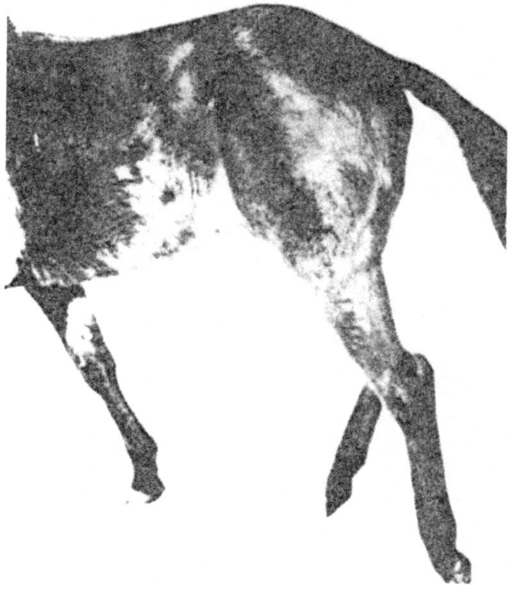

Photograph of locked stifle. Notice that the stifle and hock joints are locked in a fully extended position.

2. Treatment

- Surgical cutting of the medial patellar ligament.
- If bone cysts or patella degeneration is present, these animals have a poor prognosis.
- Arthritis occurs without treatment.

Many other conditions cause lameness in the equine. Refer to lameness text books, access the Internet for advice, or Fax consultation to various experts. If time and rest do not solve the problem, it may be best to find another animal for use.

Section 11

Diseases

Diseases

Disease may be defined as any change from the normal function of the body. The causes of disease are bacteria, viruses, protozoa, parasites, fungi, poisons, trauma, and nutritional deficiencies.

Bacteria, viruses, and parasites are living organisms which live inside or outside of the affected (host) animal. These agents can be spread by 1) biting insects, 2) direct contact with infected animals, 3) indirect contact through blood, secretions and excretions of infected animals carried by people or other objects.

Bacteria: These are single celled organisms which are individual forms of life. Given ideal conditions of warmth, nutrients, or moisture, most of them can also multiply outside the animal's body. Under adverse conditions, some bacteria can turn themselves into very resistant spores which may survive for many years.

Viruses: Viruses are smaller than bacteria and can be seen only with special microscopes. Unlike bacteria, they cannot multiply outside the animal's living cells. Once inside the animal, viruses penetrate the host's cells and use the cell's metabolic processes for their own purposes. Viruses are usually quite selective in the sites they choose to infect.

Protozoa: Protozoa belong to the animal kingdom and are larger than bacteria. In some diseases, protozoa undergo part of their reproductive cycle in an intermediate host such as a tick or fly.

Rickettsia: Rickettsia are bacteria that multiply only in living cells.

Diagnosis of Disease

Section 2 covers the normal horse basics, along with examination methods to determine if there is a disease problem. Please refer to this section to learn the proper examination procedure to determine disease conditions.

Prevention of Disease

Prevention of disease is the most cost effective method ofdisease control. It is much more economical to prevent problems than to treat them. Important factors in disease prevention include the following:

1. Nutritional status: Animals in poor physical condition due to inadequate nutrition are much more likely to become ill. Some nutritional deficiencies actually produce disease conditions. Adequate levels of vitamins and minerals will help the immune system fight disease.
2. Parasite load: Heavy parasite loads make the animal much more susceptible to disease. Some parasites actually transmit the disease.
3. Weather conditions: Extreme weather conditions of hot or cold can lead to increased susceptibility to disease.
4. Stress: Any stress that lowers the immune function makes the animal more susceptible to disease.
5. Age: Young and old animals have immune systems that do not function as well as normal adults. These animals are more susceptible to disease.
6. Sanitation: Poor levels of cleanliness can increase the incidence of disease.
7. Immune status: Some animals naturally have more immunity to disease than others. Vaccinations also increase resistance to disease.

Vaccines

Vaccination is the process of introducing certain disease organisms that have been modified so they do not produce the disease but do stimulate protection to the disease through antibody production. Antibodies are made by the body to seek out and destroy invading germs, thus giving protection against the disease. Vaccines are not available for all diseases and some are more effective than others. Vaccines must be handled carefully using refrigeration and protection from sunlight. Most vaccines need a booster to maintain protection. Refer to the information sheet that comes with the vaccine for optimum recommendations.

Respiratory System

The respiratory system includes the nose, trachea, and lungs. This system is involved with the intake of oxygen from the atmosphere and the release of carbon dioxide. The respiratory system also filters inhaled air.

Viral Infections

Influenza: one of several viruses causing flu symptoms.

1. **Signs**
   - fever, lethargy
   - dry cough
   - decreased or no appetite
   - nasal discharge
2. **Treatment**
   - rest
   - antihistamines
   - antibiotics to prevent secondary bacterial infections
3. **Prevention**
   - vaccination with booster every 3 to 6 months
   - quarantine of affected individuals

Rhinopneumonitis: a herpes viral infection causing respiratory disease, abortion, and neurologic (nerve) problems.

1. **Signs**
   - Respiratory form of this disease shows mild temperature elevation, nasal discharge, coughing.
   - Abortion form shows late term abortions or stillbirths
   - Neurologic form shows staggering, weakness in rear end, and sometimes paralysis in the hind quarters.

2. **Treatment**
   - rest, good nutritional support
   - antibiotics if signs persist
   - neurologic form may require the animal to be euthanised (humanely destroyed)
3. **Prevention**
   - vaccination every 3 to 6 months
   - Separation of sick animals from healthy animals (quarantine).

Viral Arteritis: virus disease affecting the blood vessels.

1. **Signs**
   - fever, depression, muscular weakness
   - congestion of the nasal membranes and nasal discharge
   - edema (swelling) of eyelids, abdomen, legs
   - some develop colic, diarrhea, dehydration
   - In pregnant females, abortion may occur within 7-14 days.
2. **Diagnosis**
   - signs
   - virus isolation, serum tests
3. **Treatment**
   - Supportive (make sure they are comfortable, fed, etc.)
   - relieve clinical signs
4. **Prevention**
   - quarantine
   - vaccination in some countries

Bacterial Infections

Streptococcus Equi: Also called Distemper or Strangles. Caused by a bacteria called *Strep equi.*

1. **Signs**
   - fever
   - depression, loss of appetite
   - nasal discharge

- abscesses of lymph nodes of the head & neck
- can develop internal abscesses

2. **Diagnosis**
   - bacterial culture
   - symptoms
3. **Treatment**
   - allow abscesses to "come to a head", lance, & drain
   - Antibiotics are not always needed. If used, penicillin is the drug of choice <u>but only after</u> the abscess has been drained. Treat daily for two or more weeks.
   - supportive care, anti-inflammatory drugs (p. 231), hot packs
4. **Prevention**
   - isolate affected horses
   - vaccination does not work very well for this disease

Corynebacterium or Rhodococcus: a bacterial infection most commonly found in foals.

1. **Signs**
   - lung abscesses most common
   - can cause abscesses or infection in other areas of the body
2. **Diagnosis**
   - bacterial culture
   - signs
   - Radiograph (x-ray) or ultrasound
3. **Treatment**
   - guided by culture
   - is difficult to treat in abscess form in the lungs
   - Rifampin and Erythromycin can be effective
4. **Prevention**
   - isolation
   - change foal raising area

Bacterial Pneumonia: infection of the lungs caused by various bacterial infections. These need to be cultured andtreated long-term with appropriate antibiotics. If the lung forms an abscess, the prognosis is very poor.

Fungal Infection

Aspergillis: a fungus commonly found in the guttural pouch deep in the throat.

1. **Signs**
   - nasal discharge
   - chronic bloody nose
2. **Diagnosis**
   - endoscope and culture of pouch through the nostril
   - signs
3. **Treatment**
   - long term therapy with correct anti-fungal agent
   - very poor prognosis

Other Respiratory Conditions

Bronchitis, Heaves: chronic inflammation of the bronchi (air tubes to the lungs). This can be caused by inhalant allergy, irritants, or toxins.

1. **Signs**
   - non-productive cough
   - difficult breathing, wheezing lung sounds on auscultation with a stethoscope
   - weight loss
   - nasal discharge
2. **Treatment**
   - avoid dust, mold, and irritants
   - wet hay down or put horse out on green pasture
   - steroids, such as oral prednisolone for 7 days, then give every-other-day, followed by therapy until symptoms are resolved
   - antihistamines

3. **Prevention**
   - avoid dusty or moldy feeds
   - avoid lung irritants and toxins
   - Broncho-dilators

Lungworms: see Section 8 on parasites

Nervous System

This consists of the central nervous system (brain and spinal-cord) and the peripheral nervous system (nerves coming off the spinal cord).

Viral Disease

Equine Encephalomyelitis: one of several viral diseases (Eastern, Western, Venezuelan, or West Nile) that affects the central nervous system of horses. The virus is carried and spread by mosquitoes. Affected horses are not contagious to other horses by direct contact, only by mosquito transfer of the virus.

1. **Signs**
   - elevated body temperature
   - loss of appetite
   - head pressing, leaning against an object, wobbly
   - walking in circles
   - blindness, lack of mental response
   - some animals become recumbent, laying on side, struggling with incoordination
   - death may follow
2. **Diagnosis**
   - Signs
   - paired serum samples two weeks apart sent to a lab
   - brain examination by pathologist
3. **Treatment**
   - supportive care
   - keep in shaded, well bedded area
   - Give IV fluids.

4. **Prevention**
    - vaccination annually or every 6 months in tropical climates, one month before mosquito season
    - mosquito control measures

Rabies: a virus disease of warm-blooded animals found in most parts of the world. Bats and wild animals like the fox, racoon, and skunk act as a reservoir of infection for domestic animals and people.

1. **Signs**
    - depression, mania, wobbliness
    - colic
    - ascending paralysis starting at the hindlimbs working forward
    - fever, unable to swallow
2. **Diagnosis**
    - clinical signs
    - brain examination
3. **Treatment**
    - there is no cure or treatment for this disease
    - avoid contact with these animals
    - humans are at risk with exposure
4. **Prevention**
    - annual vaccinations with approved vaccine
    - avoid contact with aggressive skunks, foxes, bats etc.
    - Do not feed carcass of animal suspected of dying from rabies to other animals—burn or bury it with disinfectants.

Bacterial Infection

Tetanus: bacterial disease of the nervous system caused by the toxin of *Clostridium tetani.* Soil is the habitat of the tetanus organism. Wounds are then contaminated with the organism where it multiplies, producing toxins that cause the symptoms. Mortality is 80%.

Tetanus—sawhorse stance.

1. **Signs**
   - stiffness, spasms of the muscles
   - exaggerated reflexes
   - may assume a "sawhorse stance"
   - Death is due to spastic paralysis of the respiratory muscles.
2. **Treatment**
   - tetanus antitoxin may help in massive doses
   - clean the offending wound
   - penicillin injections daily
   - control the spasms by keeping the affected animal in a darkened, quiet stall
   - control spasms with tranquilizers or relaxant drugs
3. **Prevention**
   - Tetanus vaccinations annually, booster if wounded
   - clean and treat all wounds
   - antibiotic treatment for all horses with wounds

Botulism: this id due to actions of the extremely potent toxins from the bacteria, *Clostridium botulinium.* The toxins are often acquired from the ingestion of rotten, spoiled or contaminated feeds.

1. **Signs**
   - Mortality is 70 to 100%.
   - Death may occur within several hours of ingestion or up to a week later.
   - shuffling, stilted gait
   - drooling, difficult swallowing, food coming out nostrils
   - Once recumbent, horses seldom survive.
2. **Treatment**
   - supportive feeding with stomach tube
   - antibiotics to prevent secondary infections
   - use of botulism antitoxin if available (is expensive)
3. **Prevention**
   - Feed good quality feed.
   - Keep dead animals out of feed.
   - Clean all wounds.

Protozoal Diseases

Equine Protozoal Myeloencephalitis: a protozoal disease infecting the spinal cord of the horse. The organism is transmitted by the opossum.

1. **Signs**
   - weak, ataxic (wobbly)
   - may become recumbent
   - often involves only one limb
   - facial paralysis
2. **Diagnosis**
   - serum blood testing in Kentucky
   - cerebral spinal fluid tap and testing
3. **Treatment**
   - sulfadimethoxine (SMZ-TMP 2-3 mg/kg) and pyrimethamine (100 mg daily) for 1 to 2 months
   - poor prognosis, variable response to treatment

Parasitic Disease

Verminous Myelitis: this is the penetration of wandering parasite larvae into the nervous system. The severity of signs depends on the number, size, and species of parasite. Signs range from chronic, mild gait impairment to peracute paralysis and death. Treatment with Ivermectin to kill the larvae and steroids to reduce inflammation in the tissues is warranted.

Other Causes of Nervous System Disease

Shivering: This is a chronic disease of unknown cause common in draft or work animals. Potential causes include trauma, hereditary influence, or secondary to viral or bacterial disease.

1. **Signs**
   - shivering of hindlimbs causing them to jerk
   - The tail may rise in jerks while the horse is backing.
   - leg jerks forward then shakes and quivers with a gradual lowering
   - increases in severity with time, resulting in muscle wasting
2. **Treatment**
   - There is no known treatment.
   - Long rest periods will help signs subside temporarily.

Wobbles: this is a sporadic, nonparalytic disease marked by incoordination. The disease is caused by a narrowing of the spinal cord canal within the neck vertebrae in young horses or due to trauma to this area.

1. **Signs**
   - lack of coordination marked by a wobbling gait
   - onset can be gradual or sudden, increasing incidence up to age two
   - dragging of toes, knuckling of fetlock are common
   - as the problem continues, the horse looses condition of the hindquarters
2. **Treatment**
   - there is no effective treatment

3. **Prevention**
   - nutritional deficiencies may play a part in the condition, balanced diets are important
   - Good levels of vitamin E are important in preventing some forms of wobbles.

Kumri: This is a wobbles caused by larval migration of a parasite called *Setaria digitala* common in the Far East. The parasite is common to cattle. Horses are infected with thelarvae by mosquitoes that have fed on cattle.

1. **Signs**
   - Signs are similar to wobbles above.
   - There may be blindness.
   - Signs gradually disappear or the animal will continue to degenerate depending on the level of infection.
2. **Treatment**
   - Diethylcarbamazine (caracide) has been successful.
   - Ivermectin will kill the larvae.
   - supportive treatment to alleviate symptoms

Moldy Corn Poisoning: This is a degeneration of the white portion of the brain from a toxin after ingestion of moldy corn.

1. **Signs**
   - tremors, weakness
   - loss of ability to swallow, staggering
   - death in 48-72 hours after onset
2. **Treatment**
   - there is no treatment
   - Prevention is the key. Avoid moldy, spoiled feeds and old bread.

Bracken Fern Poisoning: Consumption of bracken fern for 30 to 60 days will result in signs because the fern has caused a deficiency of vitamin B1.

1. **Signs**
   - weight loss, unsteadiness, swaying from side to side
   - muscle tremors, finally severe spasms and death
2. **Treatment**
   - injections of thiamine for 7-14 days
   - burn all bracken fern or alternate grazing every 3 weeks

Digestive System

The digestive tract of the horse consists of a muscular tube called the alimentary canal starting at the lips and terminating at the anus along with several organs. The alimentary canal consists of the mouth, pharynx, esophagus, stomach, small intestine, cecum, large colon, small colon, and rectum. This canal is about 72 feet long in the adult horse. The associated organs are the teeth, tongue, salivary glands, and pancreas.

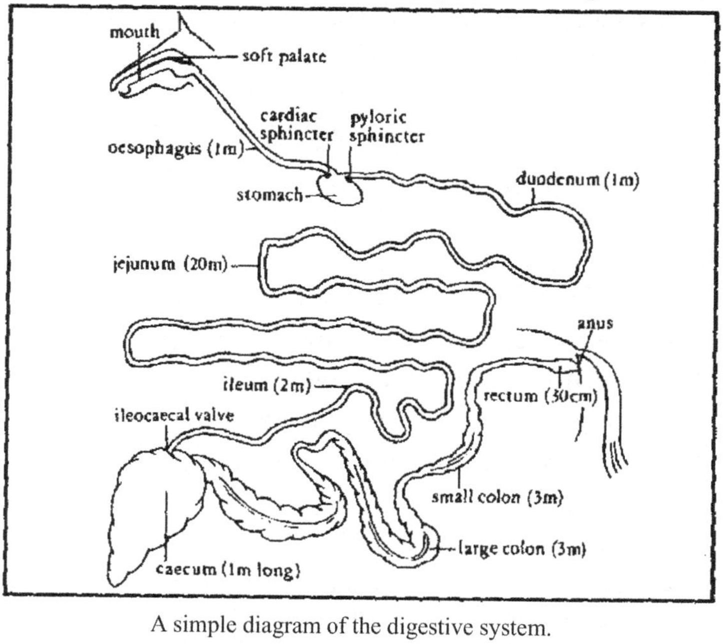

A simple diagram of the digestive system.

Dysphagia: This is difficult swallowing caused by a variety of problems including pain, mechanical obstruction, dry dirty feeds, nerve dysfunction, infectious disease, and brain damage.

1. **Signs**
    - inability to swallow
    - food and saliva in nostril
    - pneumonia may result secondarily

2. **Treatment**
    - thorough exam to determine cause and rectify if possible
    - Caution must be exercised, Rabies will show some dysphagia. **DO NOT** become exposed if you suspect Rabies.
    - Look for debris stuck in the mouth and remove.
    - Give supportive therapy of fluids and feed via stomach tube.

Choke: This is an esophageal obstruction by feed or foreign bodies that may be partial or complete. Death may result from penetration of the esophagus.

1. **Signs**
    - distressed, choking, coughing
    - feed and saliva coming out of the nostrils
    - inability to take in feed
2. **Treatment**
    - sedative to relax the horse
    - Use a smooth ended tube in the esophagus to push the obstruction into the stomach.
    - The addition of water to soften the mass may help but use a small amount so it doesn't get into the lungs.
3. **Prevention**
    - Provide soaked feed (mush form) if they tend to choke.
    - do not feed corn cobs or rough material that may lodge
    - Do not give large pills to horses; crush them first.

Colic: This term refers to any pain from the abdomen. The primary cause of pain is distention of the stomach or intestines. This distention may be caused by an accumulation of gas, fluid, or feed. The accumulation problem can be due to obstruction of the bowel or lack of gut motility to move material through the tract. Colic may range from a simple indigestion and gas accumulation to complete obstruction of the digestive tract with resultant death. Simple cases will respond to medication. Complicated cases will need surgery to save them,which is not practical in most situations.

1. Signs

- stretching, looking at abdomen, yawning
- pawing the ground, sawhorse stance
- rolling, thrashing
- decreased or absent gut sounds
- elevated heart rate
- pale or blue membranes in later stages (undersides of the lips or gums)

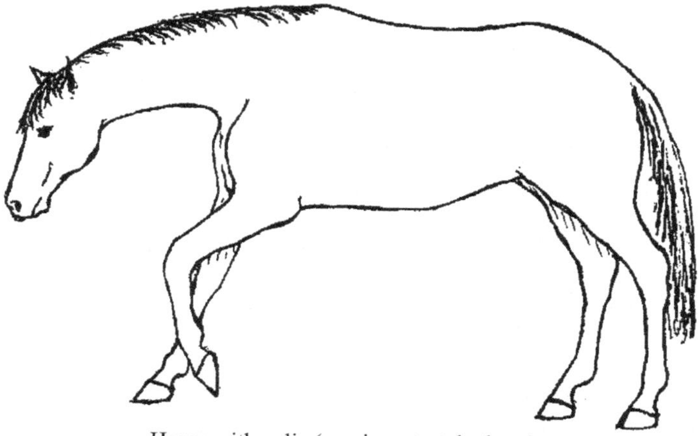

Horse with colic (pawing, stretched out).

2. Treatment

- Dipyrone or Banamine to alleviate gut spasms
- pain medication or tranquilization
- pass stomach tube to relieve gas or fluid
- mineral oil or milk of magnesia given via stomach tube if no excess fluid comes back from stomach when passing a tube

- Keep the animal warm.

- Walk the animal if it is wanting to roll.
- If the pulse is over 100 bpm, gums are pale or blue, and/or the horse is uncontrollably thrashing, the prognosis is very poor. These animals should be humanely killed if surgery is not an option.

3. Prevention

- Parasites cause up to 90% of colic. Deworming at least every 3 months to prevent parasite-caused colic (see Section 8 on Parasites)
- Avoid sudden feed changes.
- Do not give cold water to horses while they are hot from work, which causes spasmodic colic.
- Avoid spoiled or moldy feeds, which upset the gut or could contain toxins.
- Avoid feeding on sandy ground to prevent sand ingestion and impaction.
- Avoid feeding large quantities of rich feed like wheat.

Diarrhea: Diarrhea or enteritis is an inflammation of the gut and is often secondary to an infectious disease. Causes can be endotoxemia from ingestion of spoiled feeds or unsanitary conditions, parasite infection, ingestion of heavy metals, salmonella infection, improper use of some antibiotics (tetracyclines), *protozoa* (trichomoniasis), fungal infections (aspergillis), Castor Bean poisoning, bacteria, viruses, and sand ingestion.

1. Signs

- diarrhea
- abdominal pain, anxiety, depression
- dehydration
- signs can last for several days to several months depending on the cause and severity.
- death can result in some cases

2. Treatment

- Determine the cause if possible and treat accordingly e.g. fecal exam to diagnose protozoa, parasites, or sand. Examine feed sources and sanitation.

- fluid replacement by IV methods or via stomach tube. Some individuals may need 40 to 60 liters per day.
- Calcium carbonate 2-3 tablespoons in feed daily
- Give supportive care

3. Prevention

- Control parasites.
- Improve sanitation.
- Utilize good pastures, avoid poisonous plants.
- Avoid toxic substances.
- Provide good quality feed.
- Keep teeth floated.
- Immunize against infection where possible.

Skin Diseases

Parasite Larvae Infection: Various parasite larvae can infect the skin of the horse including Onchocerca, Habronema, etc.

Signs

- Along areas of the midline, chest, or legs the skin is rough and crusty with hair loss.
- Sores with itching may result.
- Ulcers can form.

Treatment

- fly control
- Ivermectin administration monthly during the fly season
- topical wound care
- see also Section 8 on Parasites

Allergic Dermatitis (Summer Itch): This is a common condition involving inflammation of the skin, which causes intense itching. The condition is the result of allergic reactions to bites from midges, mosquitoes, and flies.

1. Signs

- intense itching
- redness, scaliness, hair loss

- welts or sores on the skin
- trauma to the skin from itching

2. Treatment

- insecticides
- Keep in screened shelters in late afternoon & evening.
- steroids or antihistamines to relieve itching
- aloe or soothants for the skin
- Insecticide mixed in petroleum jelly and applied to the animal will help.

Photosensitization: This is an abnormal reaction of the skin to sunlight on white areas of the body, especially the nose and legs. This condition usually involves the ingestion of a weed (St. John's wort, Klamath weed, buckwheat etc.) containing a sensitizing agent that causes the skin reaction when exposed to sunlight.

1. Signs

- Lesions are limited to unpigmented parts of the body.
- redness, swelling, itching
- crusting areas of skin that sloughs with painful sores underneath
- This can also cause liver damage and icterus (yellow membranes).

2. Treatment

- Remove animal from access to weeds
- Put the animal in shade during the sunlight hours
- Graze only during the night
- topical protectants and soothants (aloe, AD&E cream)
- steroids orally

Mange: This is a contagious disease cause by several mites.

1. Signs

- papules (round elevations of the skin), dry crusts
- thickening of the skin, hair loss
- itching with some mites

2. **Treatment**
   - Isolate affected animal.
   - Disinfect blankets, brushes etc. between animals
   - Tricholorfon can be poured down the back of the affected horse (use 2-3 oz.) Repeat in 10-14 days.
   - Lindane dip or spray is effective.

Ringworm (Girth Itch): This is a common fungal infection of the skin. The infection is contagious and can be spread by tack and brushes in addition to direct contact.

1. **Signs**
   - round, scaly, crusty patches of hairless skin
   - There may be some itching.
   - Sores may develop.
2. **Treatment**
   - topical 10% chlorox, tamed iodines (Betadine), or lime sulfur applied daily for several weeks
   - topical thiabendazole applied daily
   - Oral griseofulven may be needed in severe cases.
   - good vitamins and nutrition

Systemic Diseases

Anthrax: This is a very contagious bacterial disease of man and animals. In the horse it is very rapid and fatal. The disease is spread by insects which carry the bacterial spores. These spores can live for long periods in the soil and contaminated feed.

1. **Signs**
   - Rapid death is most common.
   - Some cases may show swelling of the throat, neck, or abdomen along with a high fever.
   - Severe depression and stupor follow.
   - bloody discharge from body openings
2. **Treatment**
   - massive doses of penicillin for 7 days or more
   - must catch in the early stages to be successful
   - Death is the most common result.

3. **Prevention**
   - Vaccines may be available.
   - strict quarantine
   - Burn dead carcasses.
   - control flies

Caution...this disease is contagious to humans, use strict asepsis (cleanliness) when dealing with these animals!

African Horsesickness: This is a highly fatal viral disease transmitted by biting insects to horses, mules, and donkeys. The disease is common in low-lying swampy areas in hot and humid conditions. Insect-killing frost will stop an outbreak.

1. **Signs**

   A) Fever or Mild Form
   - redness of eye membranes
   - high temperature for a few days
   - labored breathing
   - Most recover from mild form.

   B) Pulmonary Form
   - rarely recover
   - fever in later stages
   - sweating, weakness, collapse
   - severe fluid congestion of respiratory tract

   C) Cardiac Form
   - edema in head and neck
   - fever, labored breathing
   - redness of eyes
   - some abdominal pain
   - Some recover but have heart damage.

   D) Mixed Form
   - combination of cardiac and pulmonary forms
   - death

2. **Diagnosis**
   - Signs may resemble EIA (Infectious Anemia).
   - serum testing at a laboratory

3. **Treatment**
   - There is no effective treatment.
   - supportive care, rest
   - Mortality rate is generally 90-95%.
4. **Prevention**
   - insect control
   - quarantine of infected animals
   - Burn the carcass of the infected animals.

Babesiasis (Piroplasmosis): This is a protozoal disease found in most tropical and subtropical parts of the world. The parasite is carried by blood-sucking ticks and infects the red blood cells of the horse.

1. **Signs**
   - pneumonia, digestive upset, kidney dysfunction
   - depression, elevated body temperature
   - anemia
   - jaundice (yellow membranes)
   - reddish brown urine
   - weakness, rapid weak pulse
   - Recovered animals can be carriers for up to one year unless treated.
2. **Diagnosis**
   - finding the parasite in the red blood cells upon blood examination by the lab
   - serology (blood work) to determine carriers
3. **Treatment**
   - oxytetracycline administration
   - imidocarb dipropionate (Imidocarb) at *2mg/kg every 24* hours for two treatments with dose divided in 4 IM sites
   - Medicine must be given in the early stages of the disease to be effective.
4. **Prevention**
   - Control ticks.
   - Isolate infected animals.
   - Treat infected animals to remove carriers or humanely kill the animal if severe and burn the carcass.

Equine Infectious Anemia (Swamp Fever): This is a viral disease carried by blood sucking insects or dirty needles from horse to horse. Horses surviving the disease often become carriers to infect other horses through the insects.

1. **Signs**
   - anemia
   - rapid weight loss, rough coat
   - fevers for 10 to 30 days or intermittent fever
2. **Diagnosis**
   - serum testing (Coggins Test)
3. **Treatment** - none
4. **Prevention**
   - Eliminate carriers by humane destruction.
   - Control insects.

Epizootic (Ulcerative) Lymphangitis: This is a rapid-spreading, usually fatal disease caused by a fungus called *Histoplasma faciminosum.* The fungus gains entry through the skin and is often carried by flies.

1. **Signs**
   - The organism affects the skin, eyes, mucous membranes, and the lungs when inhaled.
   - Abscesses develop with thick yellow oily pus.
   - In final stages, the horse looses its appetite, has a fever, and is very weak.
2. **Diagnosis**
   - laboratory examination
   - similar to other conditions
3. **Treatment** - none
4. **Prevention**
   - Destroy affected animals and burn the carcass.
   - Disinfect and quarantine contact areas for 6 months.
   - Control flies.

Glanders: This is a contagious bacterial disease that affects other animals and humans. The disease is caused by a spore forming bacteria called *pseudomonas mallei.* Glanders is found in the Middle East, the Indian subcontinent, Southeast Asia, and Africa. It is transmitted by contaminated items, direct contact, and by giving dogs infected meat to eat.

1. **Signs**
   - may be skin, lung or nasal form
   - fever, coughing, nasal discharge
   - enlargement of mandibular lymph glands (in neck)
   - Nodules on the nasal septum rupture producing ulcers.
   - With the cutaneous form (farcey) the rear legs have nodules and sores.
2. **Diagnosis**
   - association with infected horses
   - skin test similar to TB testing
   - culture the organism
3. **Treatment & Control**
   - Humanely kill the animal and burn the carcass.
   - Disinfect stables and contaminated areas.
   - antibody testing to rid carriers

Caution, this disease is very contagious to humans, use extreme care when working with suspected animals!

Tuberculosis: This is a contagious bacterial disease which can affect people. Horses are infected by the cattle bacterium.

1. **Signs**
   - weight loss in spite of normal appetite
   - dry, shaggy coat
   - stiffness
   - respiratory infection with nasal discharge
2. **Treatment & Control**
   - Humane destruction is best.
   - Burn the carcass.
   - The nasal discharge is infectious.

Trypanosomiasis: includes the diseases of Nagana, Surra, Mai de Caderas and Murrina that are caused by a protozoa. These diseases occur primarily in Africa, the Middle East, South America, and Central America. The protozoa are transmitted by tsetse flies and the vampire bat.

1. **Signs**
   - anemia
   - intermittent fever
   - edema of the lower limbs and abdomen
   - Affected horses become weak and thin.
   - Death can occur acutely or after a prolonged wasting disease.
2. **Treatment**
   - effective fly control
   - Treatment is not effective and should be avoided in cases of Dourine (trypanosoma of the genitals).
   - Drugs that are effective against the trypanosomes such as antrycide may help some of the lessor forms.

Section 12

Dental Care & Age Determination

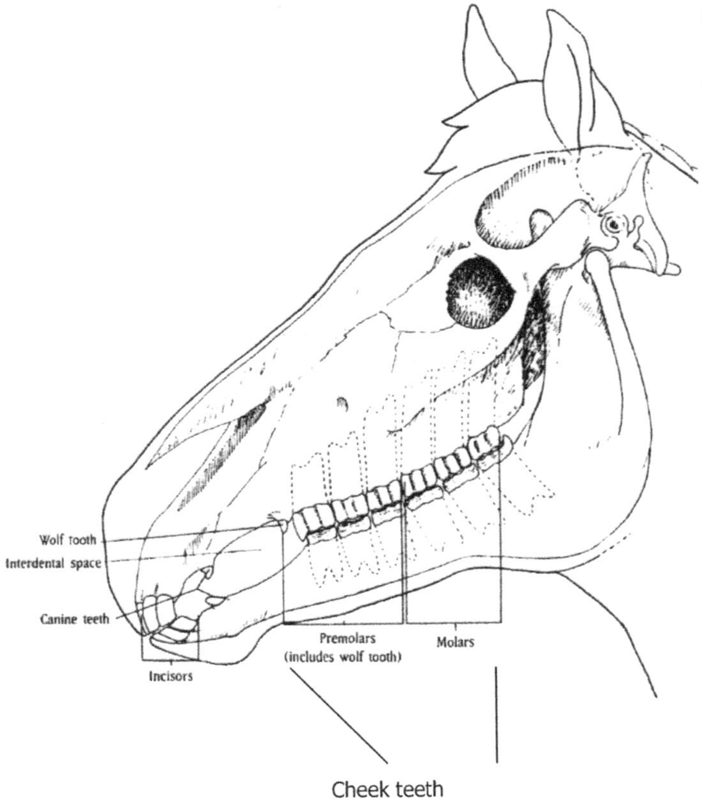

Function

Lips: The lips are selective and allow the horse to pick up objects with the mouth

Incisor Teeth: The incisor teeth are the front row of teeth on the top and bottom of the mouth. They function as a cutting apparatus for the horse. Incisors have deep enamel invaginations (clefts) partly filled with cementum. As the tooth wears, the surface has a central ring of enamel surrounding the infundibulum (center enamel cup) and a peripheral enamel ring. The changes in this surface are important to age determination (see diagram under aging).

Cheek Teeth: The cheek teeth line each side of the mouth on top and bottom and are responsible for chewing and grinding. The temporo-mandibular joint (joint in jaw) accommodates the side-to-side movements of the mandible to grind the food. These teeth have considerable reserve crownsthat erupt as the exposed crown is worn away by chewing.

Wolf Teeth: This is the first premolar. It may be absent or present in a very small form just in front of the upper cheek teeth.

Canine Teeth: These teeth are usually absent or very small in the mare. The canine teeth do not serve any purpose in the horse.

Deciduous Teeth: These are the milk or baby teeth that are replaced by permanent teeth later in life. The incisors usually erupt within the first two weeks of life. By 5 years of age all deciduous teeth are shed and the horse then has its complete permanent set of teeth (dental arcade).

Age Determination

Table 1. Tooth Eruption*

TEETH	ERUPTION
A. Deciduous:	
1st incisor	(Di 1) Birth or first week
2nd incisor	(Di 2) 4 to 6 weeks
3rd incisor	(Di 3) 6 to 9 months
Canine	(Dc)
1st premolar	(Dp 2)
2nd premolar	(Dp 3) Birth or first 2 weeks
3rd premolar	(Dp 4)
B. Permanent	
1st incisor	(I1) 2 1/2 years
2nd incisor	(I2) 3 1/2 years
3rd incisor	(I3) 4 1/2 years
Canine	(C) 4 to 5 years
1st premolar (wolf tooth)	(P1) 5 to 6 months
2nd premolar	(P2) 2 1/2 years
3rd premolar	(P3) 3 years
4th premolar	(P4) 4 years
1st molar	(M1) 9 to 12 months
2nd molar	(M2) 2 years
3rd molar	(M3) 3 1/2 to 4 years

The periods given for P3 and P4 refer to the upper teeth; lower ones may erupt about 6 months earlier.

* (From Sisson and Grossman; Anatomy of the Domestic Animals)

Changes in the Lower Incisors

Stage	*1st incisor (middle)*	*2nd incisor*	*3rd incisor (outer)*
Deciduous (baby)	6 days	6 weeks	6 months
Adult eruption	2½ years	3½ years	4½ years
Adult in wear	3 years	4 years	5 years
Loss of cup	6 years	7 years	8 years
Star appears	8 years	9 years	10 years

Age

1. Up to 8 years of age - determined by eruption or wear of incisors
2. After 8 years of age - estimate of age is made by wear and shape of incisors
3. After 15 years, the enamel cup (infundibulum) disappears and the tooth is more triangular shaped. See drawing.

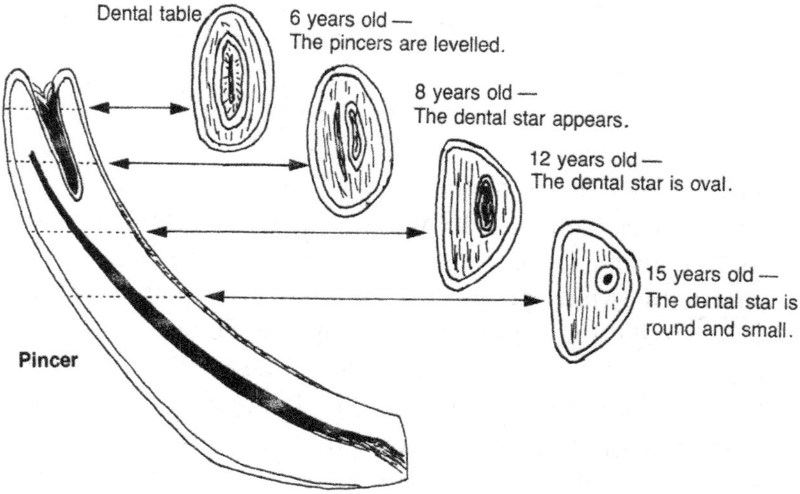

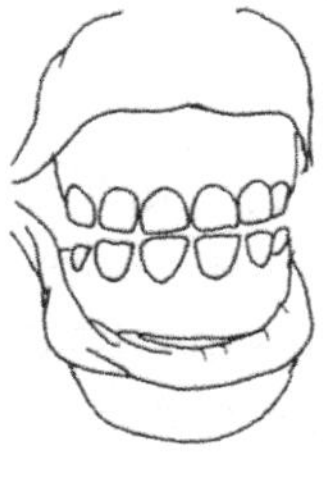

1 yr.

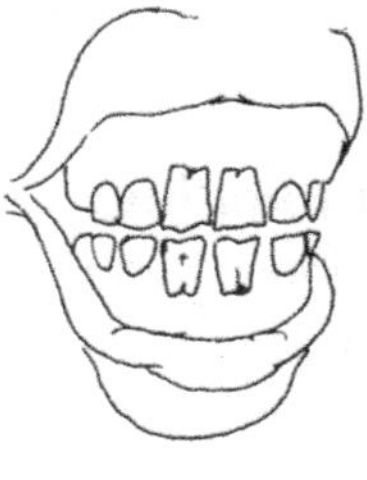

3 yr.

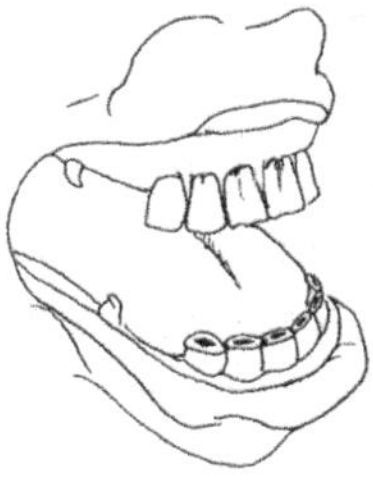

5 yr.

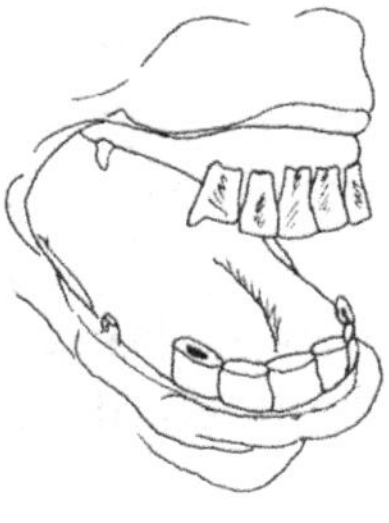

7 yr.

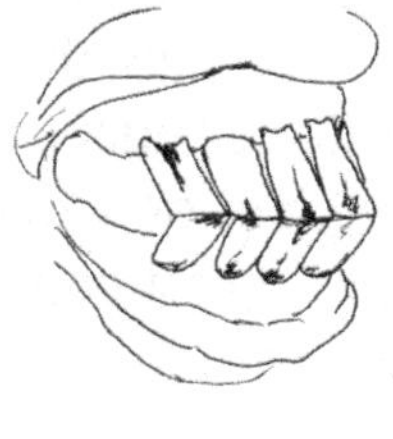

15 yr.

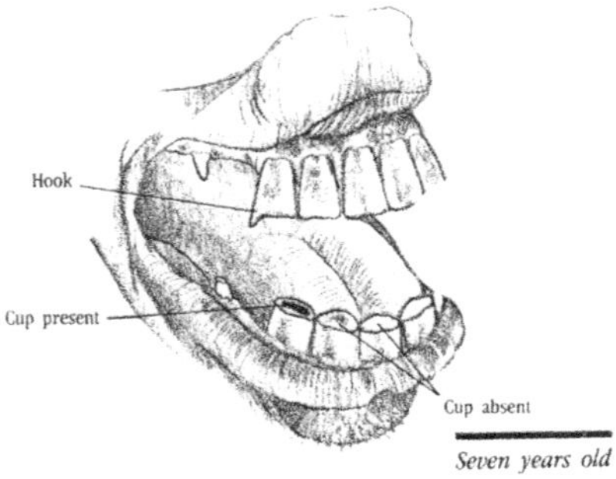

Seven years old

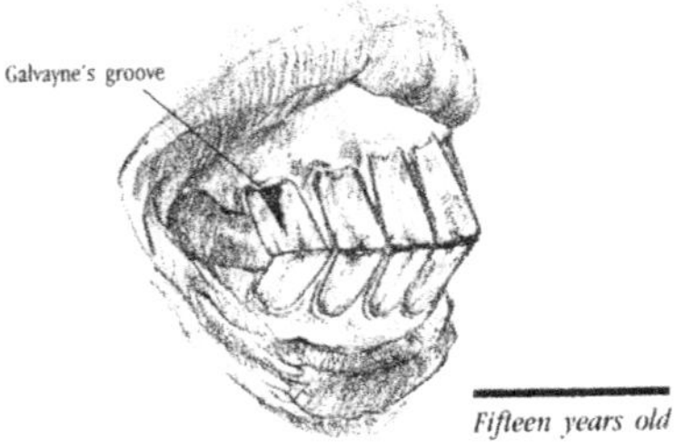

Fifteen years old

Dental Care

Wolf Teeth: these teeth are small and sit just in front of the larger cheek teeth. The bit will tend to slide up and down on this tooth causing the horse to toss it's head, especially with the snaffle bit. Many people will remove these teeth which have a very small root. Removal is accomplished by using a wolf tooth elevator or a flat screwdriver end. The instrument is wedged between the wolf tooth and the large cheek tooth. Gentle manipulation will loosen the tooth and it can be removed with needle-nose pliers. The gum may have to be cutaway from the tooth at removal.

Cheek Teeth Floating: the upper and lower arcade of teeth do not exactly oppose each other. As the horse chews, it will build sharp points of remaining tooth on the inside of thelower cheek teeth and the outside of the upper cheek teeth. A dental rasp or float is used to remove these sharp points. This removal facilitates better chewing of feed and prevents soresin the mouth resulting from cuts by the sharp points. Most horses will let you pull their tongue out one side of the mouth so the opposite side can be floated. Some animals may need tobe twitched or sedated to facilitate this procedure. Take special care when floating teeth that the horse does not strike you in the face with their front feet. Stand off to the side or use stocks if possible.

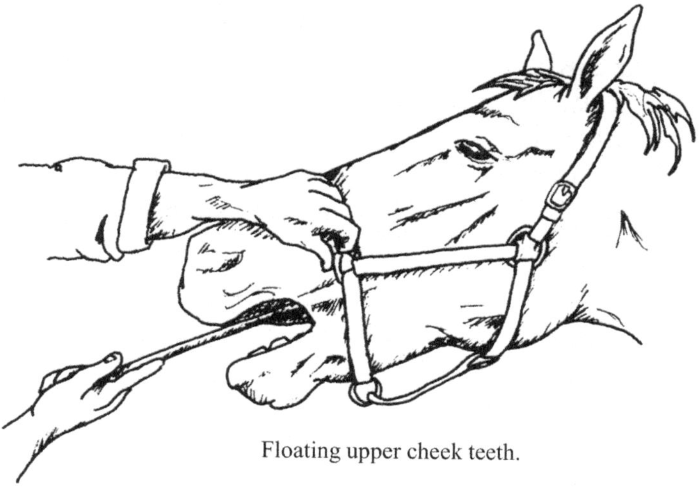

Floating upper cheek teeth.

Removal of Dental Caps: the deciduous cheek teeth sometimes stay in the mouth as flat caps on the permanent teeth. This obstruction prevents good chewing and they need to be removed. Dental caps commonly occur in pairs so the opposite side should always be examined. An ordinary screwdriver can be used to elevate the cap off the permanent tooth for removal.

Molar Spikes: the front upper and back lower cheek tooth may not sit directly opposite each other, therefore there is no tooth wear and the unopposed tooth becomes very long. As this tooth grows in length it will eventually puncture the opposing gum causing the horse pain when eating. These teeth need to be cut off. Special cutters are made for this process but will probably not be available. Using OB wire to cut the tooth is the next best process. Take care not to cut the cheek when using the wire saw to cut the spike. Float the areawhen finished. Be sure to use a flashlight to check the back tooth for spikes that need removal.

Infected Cheek Teeth: in some older horses, the cheek tooth may become abscessed with a resultant drainage out the bottom of the jaw. These teeth need to be removed by skilled surgeons with the animal under a general anesthetic. This procedure may be beyond the scope of facilities and expertise of the locale. Those animals will be thin, poor doing individuals and should not be left to suffer. Abscesses may also occur in young horses from defective teeth. Many of these animals will be in good flesh but have an obnoxious drainage and odor from the mouth.

Section 13

Equine First Aid

Whether on the trail, at work, or at home, emergencies will occur and minutes can mean the difference between life or death. If you are prepared for the worst, you can often make the best of a bad situation. Being prepared includes health and physical conditioning, a knowledge of normal body parameters, knowing what to do in case of emergencies, and possessing an emergency kit.

PREVENTIVE HEALTH CARE

Preventive health care is the most cost-effective dollar you can spend on your horse. It is much easier and cost effective to prevent an illness or disease than it is to treat that same problem. Horses and mules are needed to work and that work is diminished if we have to needlessly spend our time treatinga problem that could have been prevented.

1. **Deworming** Refer to Section 14 for parasite control.
2. **Vaccinations** Refer to Section 14 for suggestions.
3. **Conditioning**

- Condition your horse before going on long packing trips. This conditioning program will help prevent some of the muscle, hoof, and endurance problems.
- Conditioning will also toughen the skin and aid in prevention of tack sores.

4. **Hoof Care**

- Feet should be trimmed every 6 weeks and shoes reset.
- Have cracks and chipping areas treated before they become a problem.

5. **Tack**

- Check all items daily before work to be sure they are not about to break or fail in some way...refer to the equipment section (section 5).

EQUINE FIRST-AID KIT

1. **Wound Cleaning**
    mild soap for wound cleaning
    small hand brush for cleaning
    gauze sponges or clean towels
    hemostat forceps or needle nose pliers
    flush syringe
    1 oz Betadine (tamed iodine) or dilute bleach using
        2 tablespoons per liter of clean water

2. **Wound Treatment**
    wound dressing (water soluble) ointment
    telfa (non-stick) pads (6)
    6” gauze roll (2)
    cotton roll or quilted padding
    6” brown gauze (2)
    4” elastic wrap (2)
    scissors

3. **Hoof Care**
    hoof pick
    hammer
    shoe and nails

4. **Tack Sores**
    soothing ointment like aloe
    foam padding 2” thick

5. **Insect Care**
    fly spray
    small cloth to rub in ears
    insecticide mixed with petroleum jelly

6. **Assessment Tools**

thermometer in a case
stethoscope

7. **Pharmacy**

antibiotics - see Section 15, Medication Formulary
Butazoladin
Banamine or Dipyrone injection - See Section 15
ophthalmic antibiotic ointment
eye wash
sedative
12 cc syringe
18 gauge needles (15)

EMERGENCIES

Hoof Injuries

Bruises - Stones or hard ground can cause the tissues in the foot to bruise. (Refer to Sections 9 & 10)

Punctures - Punctures are very dangerous because they are often made by a small object like a nail. The object carries manure and soil into the sole where severe infections can incubate. Because the hole often seals over or is too small for adequate drainage, severe infections can result. Deep penetrating objects can also damage or infect tissues below the sole like the flexor tendons or the coffin bone. If these structures are involved, extensive measures are needed if the horse is to be saved.

1. Signs

- limping on a limb
- drainage from the sole or from the coronary band
- Affected horses also show pain when pressure is applied to the sole in the area of the infection.

2. Treatment

- Examine the sole for nails or wood splinters and remove.
- Proper debridement of the wound is important for success. This removes pus, bacteria and toxic chemicals. Open a hole for drainage into the hoof using a sharp, clean, hoof knife…this will bleed (see diagram below).
- Soak the foot in Magsulfate Salts or Epsom Salts (2 kilos per 4 liters of water) daily for 20 minutes.
- Apply iodine to the wound after soaking.
- Wrap the foot with duct tape or use a special boot to keep the foot clean until the hole fills in and hardens.
- Antibiotics should be given daily…refer to Medication Formulary (Section 16).
- Tetanus vaccination should be given.
- Contrast x-rays may be needed if the object went deep into the foot. This will tell if deep structures are damaged.

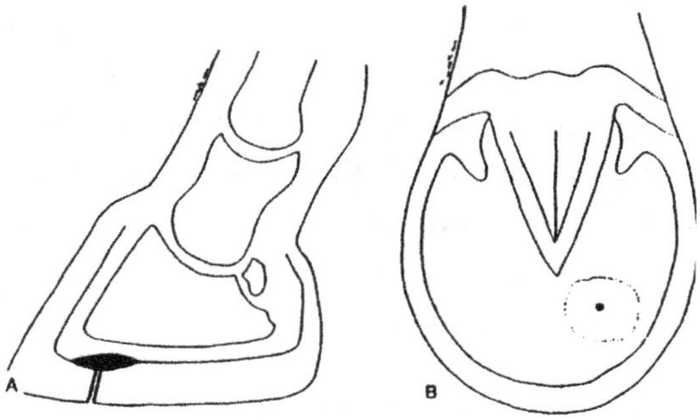

A, Side view of typical superficial subsolar abscess. A small tract communicates through the hard solar horn. B, Solar view of same subsolar abscess as in A. A small black spot in the external solar horn is typical. Deep paring over the dark spot exposes the abscess cavity (stippled area).

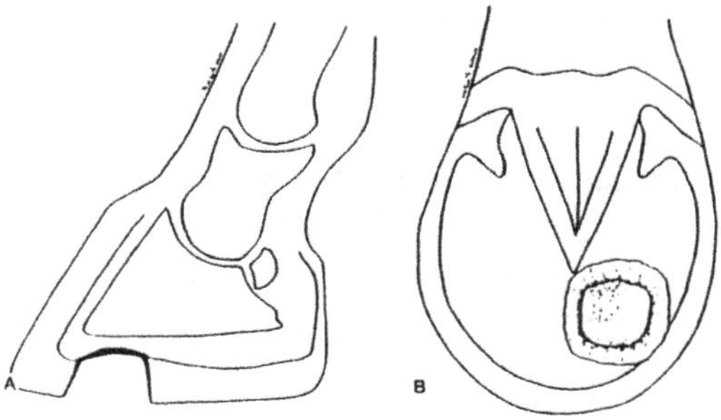

A, Side view of proper excavation of subsolar abscess. The margins of the cavity are pared until confluent with the surrounding healthy hard horn. New epilhelium *(heavy dark line)* fills the base of the cavity. B, Solar view of proper excavation of subsolar abscess. The edges of the surrounding horn are tapered down to the base of the cavity.

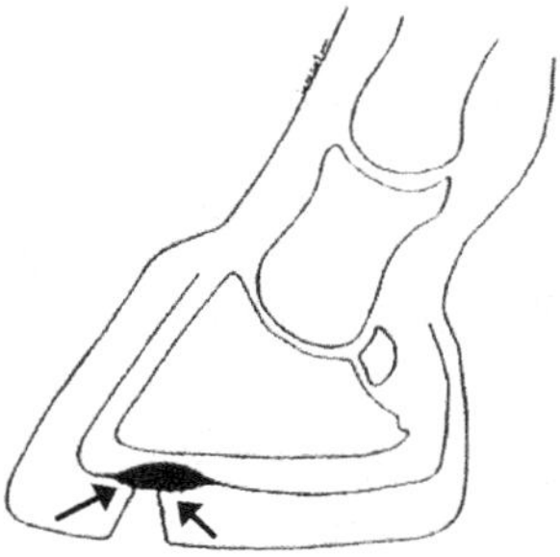

Side view of typical superficial subsolar abscess improperly drained. Shelves of hard horn *(arrows)* retard drainage.

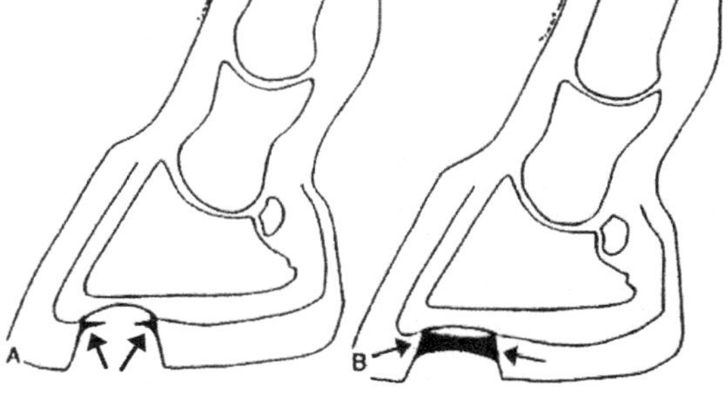

A, Side view of early healing in deeper subsolar abscess. Granulation tissue *(stippled area)* fills the wound cavity. Early epithelialization has begun *(arrows)*. B, Side view of advanced healing of deep subsolar abscess. Granulation tissue (stippled area) has decreased in quantity. New horn has thickened and filled the defect (arrows).

Lost shoe - If you take it easy on the trail and if your horse does not have sensitive and soft soles, you may be able to bring the animal home without another shoe being applied. If the horse becomes sore without the shoe another option would be to use an Easy Boot (if available) to protect the sole from rocks etc. Foam padding and duct tape can also help protect the sole for short distances.

Tack Sores - Poor fitting tack and/or soft-skinned horses can result in rub sores. These sores can be mere red spots or they can extend through the skin. The sores cause pain much like a severe blister would in your own boot.

1. **Signs**

- Check the horse over at the end of the day for red areas or sores.
- During the ride or in work, horses that are sore may exhibit discomfort, an altered gait, or even lay down.

2. **Treatment**

- Trim the hair around the area.
- Apply aloe or a soothing ointment if the wound is superficial. Use an antibiotic ointment if the wound extends through the skin.
- Take a section of foam and cut a hole slightly larger than the wound. Place the foam pad under the offending tack to protect the wound.

3. **Prevention**

- Be sure tack and saddles fit properly.
- Check your tack at rest stops on the trail.

Eye Injuries

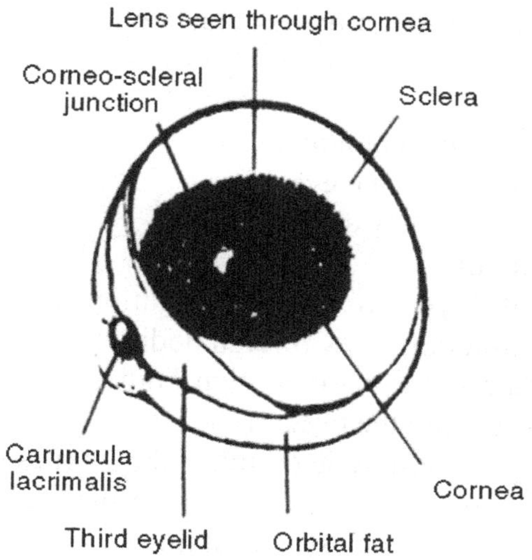

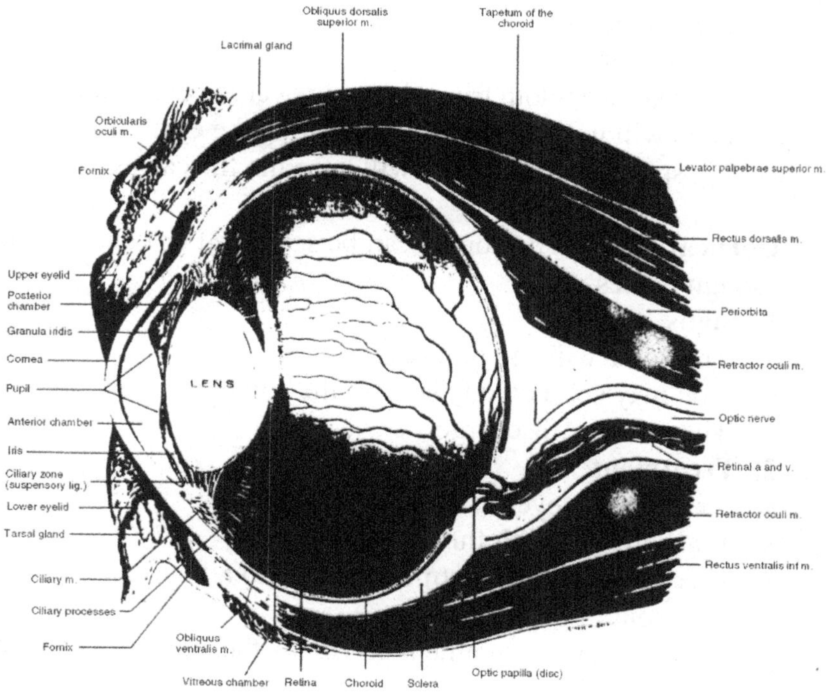

Conjunctivitis - This is an inflammation of the white tissue on the inside of the lids. Often this condition is caused by dust, flies, or other foreign bodies that produce irritation to the eye tissues.

1. **Signs**

- redness of the conjunctiva
- discharge from the eye
- The eye lids may also swell.

2. **Treatment**

- Flush the eye with eye wash solution, saline, or clean water to remove any foreign bodies and debris.
- Examine the eye for foreign bodies; remove them unless they stick into the eyeball.
- Use an eye antibiotic ointment three times daily.
- Many foreign bodies will scratch the cornea and cause an ulcer. These animals will squint—see below under ulcers.
- Wipe fly spray around the eyes to prevent further damage by flies, or mix insecticide with petroleum jelly and apply around the eye.

3. **Prevention**

- Eye masks or cloth fringe hanging over the eyes from a brow band will prevent flies from irritating the eye.
- Fly spray around the eye will keep the flies away.

Corneal Ulcers The cornea can be scratched or punctured by foreign bodies.

1. **Signs**

- The horse will squint its eyelids.
- You may see a cloudy spot on the cornea.
- The eye may discharge.

2. **Treatment**

- Examine the eye for foreign bodies.
- Flush the eye with flush solution.
- Use antibiotic eye ointment every 6 to 8 hours.
- Keep out of bright sunlight.
- Apply atropine ointment to dilate the pupil and prevent uveitis (inflammation of the inner part of the eye).

Foreign Bodies - Foxtails, dirt, sawdust, hay, pine needles etc. – anything that can lodge in the tissues around the eye or hide under the eyelids.

1. Signs

- The horse may squint.
- The eyelids may swell.
- Severe tearing or pus discharge is common.
- You may see the foreign body.

2. Treatment

- Examine the eye and remove any foreign bodies unless they extend into the cornea.
- Flush the eye with flush solution.
- Use eye antibiotic ointment every 6 to 8 hours.
- **If a foreign body extends into the eye bulb, do notremove it, seek veterinary assistance (if available) as soon as possible!** If help is not available, the foreign body must be removed. The animal may loose the use of this eye. Antibiotics, eye ointment, fly spray around the eye, tetanus, and eye protection will be needed.

Eyelid Lacerations - The eyelids can be easily torn or cut by brush and obstacles.

1. Signs

- The signs are obvious - a torn lid, often with bleeding.

2. Treatment

- These wounds heal well when sutured.
- They do not always need to be sutured immediately.
- Wash the eye out often with flush solution.
- Use eye antibiotic every 6 to 8 hours.
- Start Penicillin G or another antibiotic.
- Suture if possible.

Snake Bites - Most snake bites occur on the nose or limbs because the horse is curious and is sniffing the snake. Most Rattle Snake venom is not a problem for horses due to a small dose for the animal's size. The real problem is the bacteria that the snake carries on its fangs, which can cause massive tissue necrosis (tissue death).

1. **Signs**

- Massive swelling occurs around the site of the strike.
- Skin at the point of the point of the bite turns black and dies.
- Secondary bacterial infections can spread down tissue levels resulting in massive tissue destruction .
- The horse will get depressed from the bacterial toxins.
- Elevated body temperature.
- Pain and heat at the affected site.

2. **Treatment**

- Tetanus toxoid vaccination well before the bite occurs is very important. Then administer a booster after a bite.
- Tetanus antitoxin and clostridial antitoxins may be of benefit.
- Pen G (large doses) 25 cc twice daily or more for a 500 kg animal
- Butazoladin to reduce swelling and discomfort
- cold water hydro massage by standing in a stream
- Seek veterinary assistance as soon as possible.

3. **Prevention**

- Have current tetanus immunizations.
- Avoid snakes as much as possible.

Insects - Refer to Section 8 on Parasites.

Porcupines - Occasionally horses will walk up and sniff a porcupine. The end result will be quills in the nose or legs. These quills can cause problems if broken off under the skin or infection from the punctures.

1. **Treatment**

- Grasp the quill at the base next to the skin with a hemostat or needle-nose pliers and pull the quill out.
- If there are many quills your horse may need to be twitched or sedated during removal.
- Quills broken off under the skin may need surgical removal.
- Penicillin G should be given if there are many quills or if some are broken off under the skin.

Foxtails - Foxtails and various grass awns are usually a problem when they are brown and eaten by the horse. These awns will often imbed in the gums and under the tongue causing pain, swelling, and discomfort.

1. **Signs**

- drooling, swelling, foul breath odor
- poor appetite
- An exam of the mouth and gums will reveal the foxtails.

2. **Treatment**

- Some horses may need sedation for removal of foxtails.
- Use your hemostat forceps to grasp and remove the awns.
- Penicillin G or another antibiotic should be given.
- Butazoladin will help the swelling.
- Tetanus should have been given within 3 months prior. Booster if necessary.

Exercise Intolerance - This problem is evidenced by the horse's inability or refusal to go any further on the trail or in the arena. Pushing the horse at this point can cause severe and perhaps life threatening problems. There are several forms of exercise intolerance. Refer to Section X for further information.

Wounds

Wound Management

Step One - Inspection of the patient

- With blood loss or shock we see elevated heart rates (greater than 60 heart beats per minute), a CRT (capillary refill time) of 4 seconds or greater, cool extremities and shivering. These patients need more help than simple wound care (blankets, IV fluids, shock Rx, and pressure wraps) to stop the bleeding. Take what steps you can and seek help.
- If the patient has normal parameters, then proceed to step two.

Step Two - Inspection of the wound

- Cuts not extending all the way through the skin are rarely severe. Topical aloe may be all that is needed after the area is cleaned if the patient does not show any lameness.
- Wounds that go completely through the skin layer buthave not damaged tissue in other layers may heal without complications if they are above the knee or hock.
- Wounds below the hock or knee, damaging tissues below the skin layer or near a joint or tendon need professional assistance as soon as possible.

3. Step Three - Wound Cleaning

Horses live in dirt and manure. Therefore their wounds are often filled with this material. High bacteria loads cause infection but soil also makes infections worse. **Wound clean-ing is the most important step!**

- Use a solution of clean water and Betadine (use enough povadine iodine—Betadine—to make the solution the color of tea) and flush the wound with pressure through a syringe. Pick any debris out of the wound. Multiple flushings are very important to remove bacteria and dirt.
- If you have a bleeding vessel, clamp the hemostat on the vessel and twist it 10 to 15 times. Often this will stop the hemorrhage. If this does not work, a pressure wrap will assist until you get to a veterinarian.

4. Step Four - Medicating the wound

- Blot the area around the wound dry.
- Apply a wound ointment to the area.
- Do not use: Hydrogen peroxide, Strong soaps, Household detergents, Tincture of iodine, Creolin, Grease, Petroleum products, Talcum powders, Bleach (use only if diluted)

5. Step Five - The Bandage

If you are not good at bandaging, do not use a wrap. No wrap is better than a bad wrap. A bandage will protect lower leg wounds from dirt, keep the wound fresh for possible suturing, and help keep the wound edges together. Upper body andhead wounds seldom need wraps.

A) First (inner) Layer

- Purpose: to maintain a sterile wound environment.
- Parts: a) Topical Rx b) Telfa pad c) 6" gauze roll
- Application: Apply ointment then the telfa pad. Secure the pad with the gauze dressing. With a gentle motion spiral the wrap down the leg from the injury, then upward past the injury and finally down the leg again. Keep pressure even and overlap each successive turn so that it covers half of the previous turn (see pictures below).

B) Second (middle) Layer

- Purpose: to absorb drainage and to prevent the bandage from becoming too tight, while still supporting the limb.
- Parts: a) sheet cotton or quilted leg wrap
  b) 6" brown gauze roll
- Application - Apply the padding in a circle around the limb. Have the bottom of the padding sit at the level of the pastern. Secure this padding with brown gauze in a similar manner as you did the gauze wrap. If the padding does not extend past the wound, stack a secondcircle (layer) of padding higher up the leg.

C) Third (outer) Layer

- Purpose - to secure the primary and secondary layers and to keep the wound clean.
- Parts: a)Vetwrap or Polo wrap...some kind of elastic wrap
  b)1" adhesive tape
- Application: Start just above the bottom of the padding, spiral the wrap up the limb firmly overlapping 1/2 the material over the previous turn.

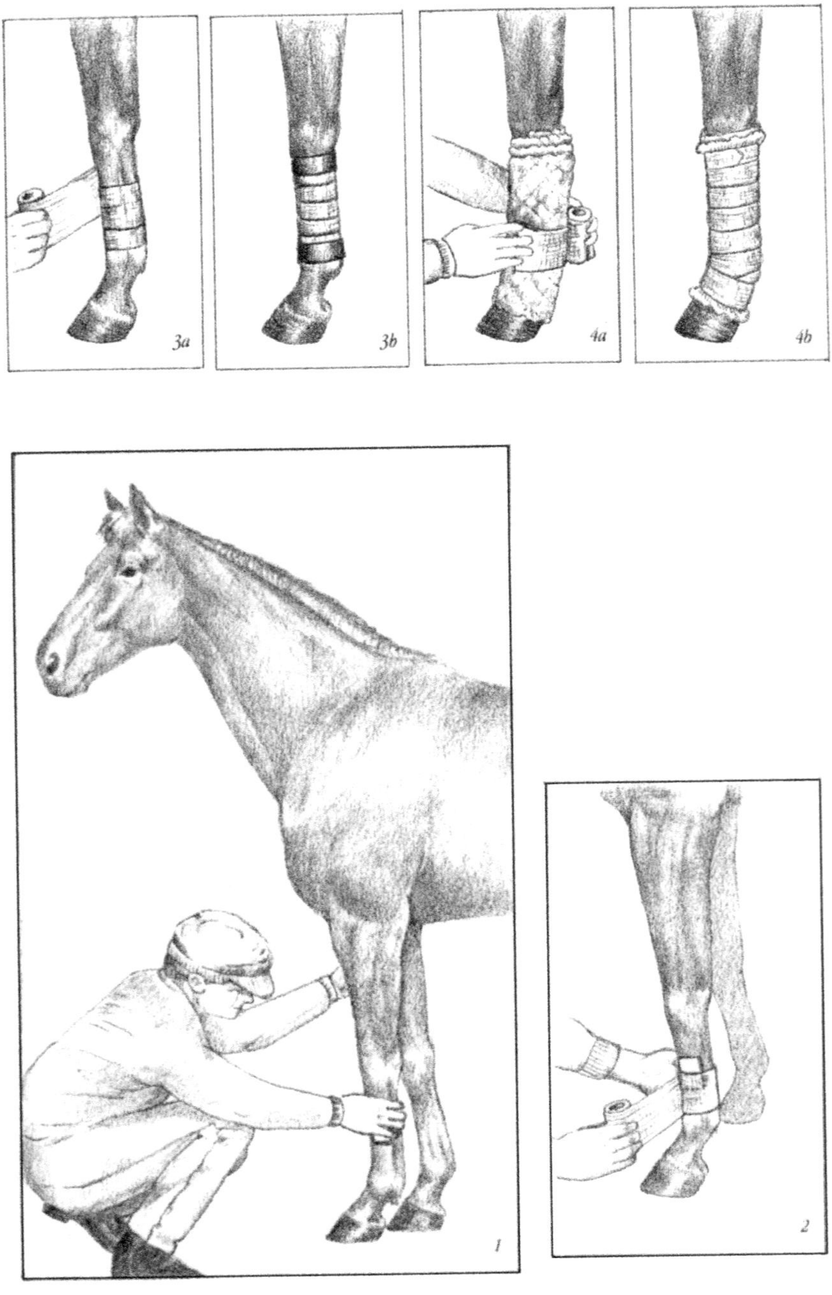
3a
3b
4a
4b
1
2

D) Splints: Wood slats or branches taped to the outside of the padded wrap will help add further stability if needed for injuries.

E) Heel Blocks: If the flexor tendon is swollen, a wedge block or round branch section taped to the heel will relieve pressure off the flexor tendon and may prevent further damage until help is available.

F) Foot Bandage - Use the same principles as in the first and possibly the second steps above. Secure these layers with duct tape or use an Easy boot. Also refer to the drawing under Hoof Care (Section 9).

6. Bandage Care

- Confine the animal to a small, clean area.
- Change the wrap daily if there is a lot of pus or exudates — otherwise change it every 2 to 3 days.
- Clean the wound thoroughly before applying a new wrap.
- Change the wrap if it slips, gets wet, smells, is swollen above or below the wrap, or if the horse seems to be in pain from the wrap.

Types of wounds

Punctures - These wounds often have massive contamination and possible tissue damage under the skin where only a small hole exists. Damaged tissue is an excellent environment for bacteria growth.

Burns - these are usually caused by ropes or from tack. The friction results in thermal damage to the tissue. This damaged tissue dies and must be removed by nature with drainage or by surgery. Once the dead tissue is removed/comes off treat the wound as an open laceration.

Lacerations (cuts or deep scrapes) - these come in all shapes and forms. Small cuts often heal with minimal treatment but crushing wounds, large open wounds, and those of the lower limbs need help to minimize scar tissue.

Antibiotics

These are needed if the wound is large, contaminated, deep, already infected, or near a joint or tendon sheath. Consult the Medication Formulary, Section 16 for antibiotic type, dose, and method of administration. The most common problems are from using the wrong drug, or giving a dose that is incorrect.

Tetanus Prevention

If your horse has not received a tetanus immunization in the last 30 days a booster is needed.

Red Alerts - Get Help!

1. You discover a wound but your horse is cold, has pale membranes, seems very tired, is breathing irregular etc. The horse may be in shock. Cover the horse with warm blankets and get help.
2. There is uncontrollable bleeding from a wound even with direct pressure for over 10 minutes.
3. The horse is very lame or is thrashing about because of the injury.
4. The wound extends deeper than the skin.
5. There is clear yellow fluid coming from the wound.
6. You find a foreign body like a wood stick in the wound.
7. The wound is below the hock or knee or near a joint.
8. As a wound heals, a deep hole persists producing pus.
9. The wound is hot, swollen, red, and draining pus.
10. The wound has filled with a large mass of red tissue.
11. The wound is not healing.

Factors slowing wound healing

1. Patient is in poor condition
2. Poor blood supply to the area
3. Steroids have been given to the animal
4. Infection in the wound
5. Foreign bodies are present in the wound
6. Motion at the site of the wound
7. Low oxygen levels to the wound
8. Improper bandages

Colic - Refer to Section 11, Diseases

Leg and Joint Problems - Refer to Section 10, Lameness

Section 14

Surgery

The surgery chapter is written by, Dr. Jack Amen. The procedures featured in this chapter are intended for use in developing countries where there is little veterinary service available. The procedures are not meant to be used by horse owners in those countries where competent veterinary is available. Some of the described procedures can be dangerous to perform by untrained personnel. Serious injury could result to the handlers and/or to the animals. These procedures should not be attempted in developing countries without skilled personnel and proper equipment available.

At various times and in some remote areas veterinarians or trained animal health providers are not and will not be available. In such situations a lay person may be called upon to perform minor surgical procedures on horses. This chapter will describe a number of procedures that lay people can and have performed on horses.

INSTRUMENTS

To perform minor surgical procedures one needs to be equipped with a few basic instruments. A minor surgery kit should include, but not be limited to, the following:

Tissue forceps

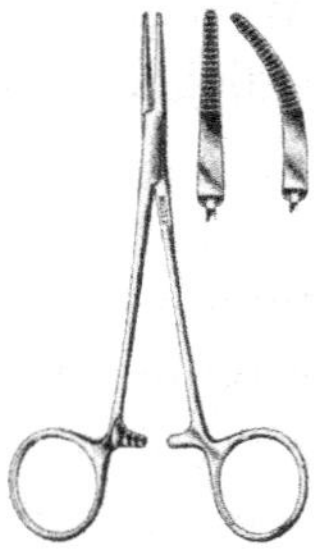

Hemostats (2 to 4 pair)

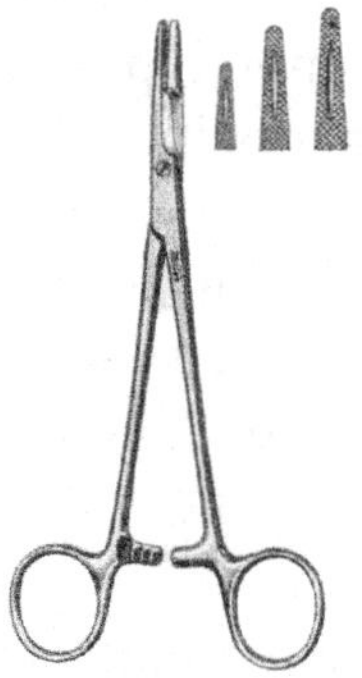

Suture needle holder

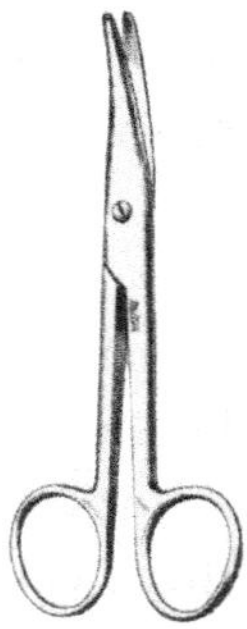

Surgical scissors (2 Pair)

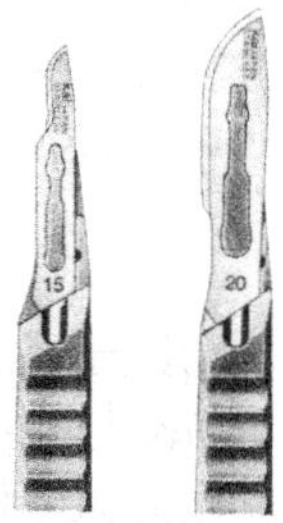

Scalpel or sharp knife

These instruments come in a variety of sizes. For minor surgeries, performed in the field, most any size will do,except when it comes to hemostats. It is easier to find and "pinch off" bleeding vessels with small "mosquito" hemostats rather than larger ones.

As stated previously, it is wise to have the proper instruments. However, instruments may not be available in an emergency situation. At this point one needs to improvisewith what is available. A successful caesarean section can be performed on a cow with nothing more than a knife blade and a pair of pliers.

SUTURE NEEDLES

Needles come in a variety of shapes and sizes. The two most common types of suture needles are the cutting point and the tapper point.

A cutting point needle is used to place sutures through the skin. The shape of a cutting point needle on cross section isa triangle. The three cutting edges of the triangle make placement of the suture material easier.

Tapper point needles are used to place sutures in muscle, tendons, internal organs, etc. The shape of a tapper point needle on cross section is round. This type of needle will not cut small blood vessels and capillaries like a cutting needle will. It is not impossible to suture skin with a tapper point needle, however more force is required to push the needle through the skin.

In this chapter we will discuss only surgeries that require a cutting needle. To suture lacerations and incisions on the body and limbs of a horse one needs to have a curved needle that is approximately three inches long. A smaller needle is used for the head, especially near the eyes and eye lids.

SUTURE MATERIAL

There are two basic classifications of suture material. They are absorbable or nonabsorbable, and monofilament or multifilament.

Absorbable sutures are just that. They will be broken down by the body and absorbed. This type of suture is usually used internally. Nonabsorbable sutures are sutures that need to be removed once the incision or laceration has healed. This type of suture is usually used in the skin.

Monofilament suture is one solid strand of fiber, such as nylon. Multifilament suture is made up of more than one strand of fiber, much like a cable or a braided rope.

Suture material also comes in a variety of sizes, varying from quite thin to quite thick. These sizes range from 7-0 through 3. 7-0 being 0.025 millimeters in diameter and 3 being 0.673 millimeters in diameter.

Suture material can be purchased in spools without needles or in small packets with the needles attached already.

Number 2 suture material is the preferred size for suturing incisions and lacerations on horses. In addition the material should be non-absorbable and monofilament. Examples would be number 2 prolene or number 2 nylon.

If number 2 prolene or nylon suture material is not available, 20 lb. to 25 lb. test nylon fishing line can be used to suture incisions and lacerations on horses. If nylon fishing line is used it should be new and soaked in a disinfectant, such as betadine (povidone-iodine) or nolvasan (chlorhexidine) before use.

INSTRUMENT STERILIZATION

Ideally the surgical instruments should be sterilized before use, in an autoclave. If an autoclave is not available the instruments can be sterilized in a pressure cooker. The time required in a pressure cooker, once it has come to a boil, is 15 minutes.

Another method of sterilization is called cold sterilization. In this method clean instruments are placed in a tray of disinfectant such as a povidone iodine (Betadine) or chlorhexi-dine solution. The disinfectant should be diluted to the proper strength; when using povidone iodine products the solution should have the coloring of tea. Any darker than that is too concentrated and will damage the tissues. The properly dilutedsolution can be used to clean and flush wounds as well.

SUTURING LACERATIONS

Any full thickness skin laceration, greater than ½ inch in length, will benefit by being sutured. Sutured lacerations will heal quicker than nonsutured lacerations and are less likely to become contaminated or infected.

An attempt should be made to suture all wounds and lacerations below the knee and hock. The reason being that untreated wounds and lacerations below the knee and hock often develop "proud flesh" or exuberant granulation tissue (this looks like a large mass of scar tissue).

The following type of wounds should not be sutured:
-Wounds that are more than 12 to 24 hours old;
-Wounds that are highly contaminated with dirt and debris
-Puncture wounds, ie. nail wounds.

The following steps should be taken when suturing any wound or laceration:

1) Give the horse a complete physical examination to rule out other problems and shock (see Section 13 on first aid).

2) The horse should be current on its tetanus vaccinations. If not, give the horse a tetanus booster. If the horse has never had a tetanus vaccination the horse should receive a tetanus vaccination and tetanus anti-toxin at the same time, one being administered on the left side of the body and the other on the right side of the body.

3) The horse should receive a sedative such as xylazine (see Section 16 on medications). This makes working on the horse much easier. If a sedative is not available a nose twitch will help.

4) The hair should be clipped or shaved around the surgery site.

5) The wound or laceration should be cleaned with a mild soap (see Section 13 on first aid).

6) Once the wound has been cleaned it should be flushed with large amounts of fluids, such as a povidone iodine solution under pressure. A large syringe, 30 to 60 milliliters in size, can be used to squirt the fluid into the wound to dislodge remaining hair, dirt, debris, and bacteria. If flush fluid is not available clean water is better than nothing.

7) The margins (edges) of the wound should be infiltrated (injected) with a local anesthetic such as lidocaine. A 1 ½ inch, 20 gauge needle is used to inject the anesthetic under the skin. Enough is injected to produce a small bump in the skin all the way around the wound. This will require inserting the needle a number of times. Needles should always be inserted

through an area that has already been infiltrated. With each insertion of the needle a 1½ inch of wound margin can be infiltrated. Again, if a local anesthetic agent is not available a nose twitch will help.

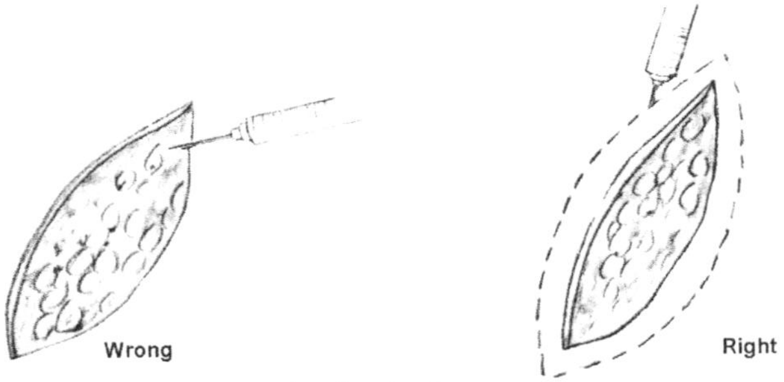

Local analgesia

8)At this point suturing of the wound can begin. Using number 2 prolene or nylon with a cutting needle, place the sutures about ½ centimeter apart. The needle should enter the skin on one side of the wound and exit the skin on the oppo- site side of the wound about ½ centimeter from the edges of the wound.

Sutures should be pulled only tight enough for the edges of the wound to touch each other. If the sutures are too tight it may lead to increased irritation, suture line swelling, and difficulty in removing the sutures after wound healing.

Three to four square knots should be tied on each suture. The suture ends should be cut, with a scissors, to a length of one to two centimeters.

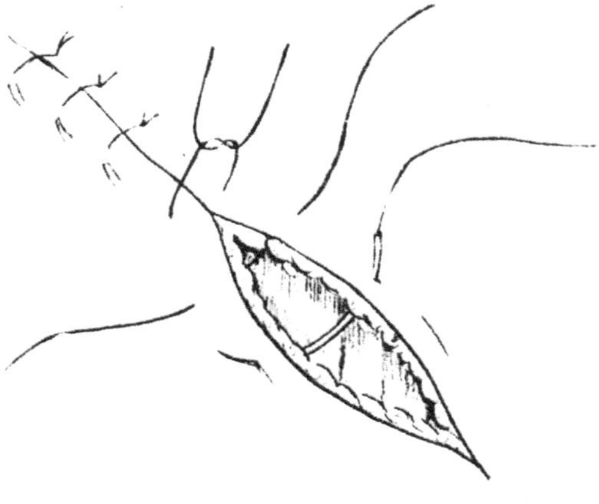

Placing simple interrupted sutures.

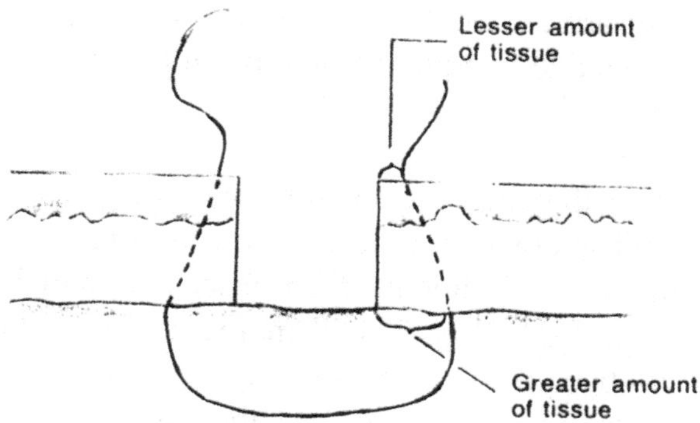

A

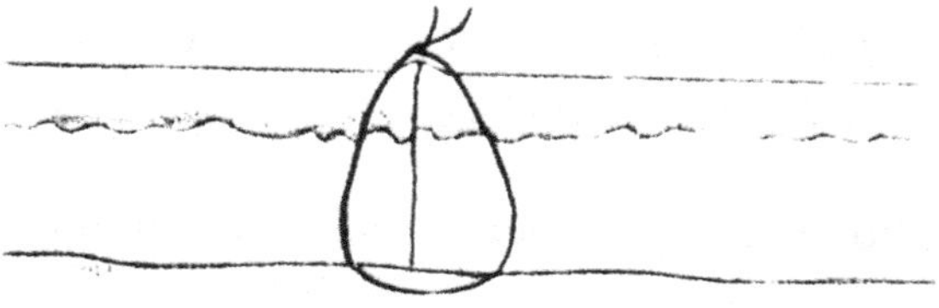
B

9) Once sutured, no additional salves or ointments are necessary. The suture line will heal better if left clean and dry.

10) Because of the possibility of proud flesh below the knee and hock, once sutured, wounds in these areas should be bandaged after being sutured (see Section 13 on first aid).

11) The horse should receive antibiotics, to prevent infection, for five to seven days. In addition, care should be taken to prevent screw worm infestation in areas where screw worms exist. This can be done by the use of insect repellants and insecticides—but do NOT use directly on the wound!

12) The sutures can be removed approximately two weeks after the surgery.

ABSCESSES

At times injury, foreign bodies and/or bacterial infection in horses will cause abscesses to develop. One common cause of abscess formation in the horse is Streptococcus equi, the

the causative agent of equine distemper or "strangles". Abscesses due to equine distemper form in the lymph nodes of the head and neck.

Whatever the cause of the abscess, one needs to wait until the abscess has come to a head before it is lanced and treated. In the initial development of an abscess it will be a small firm lump. As it develops it will increase in size, become painful, and will become warm to the touch due to inflammation. In the final stage of development the abscess will develop a soft spot in the center. This is the stage when an abscess should belanced and treated. If left untreated it will rupture on its own, but usually in the wrong place.

The following steps should be taken when treating an abscess.

Steps 1 through 4 are the same as described above in suturing a laceration.

5) The area around the abscess should be cleaned with a mild soap.

6) At this point an incision is made into the lowest point of the abscess. This should be a vertical incision of 1 to 2 inches in length. Once the incision has been made the contents of the abscess will drain out. If the incision is made too high, pus will pool in the bottom of the abscess.

7) Pressure should be applied to the abscess to squeeze out the remaining contents.

8) The abscess should be explored with an instrument or finger to make sure that all contents have been removed and that a foreign body is not left inside.

9) The abscess should be flushed with large amounts of flush fluid, such as povidone iodine, using a large syringe.

10) Antibiotics should be given to the horse for five to seven days. In addition, care should be take to prevent screw worm infestation.

11) Over the next one to two weeks the incision should be kept open and the abscess flushed every day. This should be continued until the drainage stops or the volume of the abscess pocket has sufficiently reduced in size.

CASTRATION

Castration of the male horse is usually performed at about 12 months of age. Castration at this age (instead of when the horse is younger) allows the development of muscle and bone and other desirable physical characteristics. Horses may be castrated at younger and older ages as well.

Prior to castration the following should be asked:
-Is the horse healthy?
-Have both testicles descended?
-Is the horse up-to-date on a vaccination program, especially tetanus?
-Is the horse up-to-date on an internal parasite control program? A heavy parasite load may lead to excessive bleeding at the time of castration.

INSTRUMENTS FOR CASTRATION

The instruments that will be needed to castrate a horse are as follows:
-Scalpel or sharp knife
-Emasculator (see picture on page 220)
-Scissors

ANESTHESIA AND RESTRAINT

Chemical restraint is the recommended method of restraint when castrating a horse. It is safer for humans and horses. If sedatives and other anesthetic agents are not available physical restraint may be used. The following are methods of restraint that may be used in the field:

1) First the horse is sedated with an agent such as xylazine given intravenously (see Section18 on medications). Once the sedative has taken affect a general anesthetic agent such as ketamine or thiamylal sodium is administered intravenously. At this point the horse will lie down. Care must be taken so that the horse does not fall on any harmful objects or strike its head on the ground. Once the horse is down, the top hind leg is pulled forward by an assistant. This gives better exposureand access to the scrotum.

2) In the absence of a general anesthetic agent the horse may be castrated while standing. Again the horse is sedated. While standing 10 milliliters of a local anesthetic agent are infiltrated into the skin on a line one centimeter from the center of the scrotum. This is done on both sides of the scrotum. The spermatic cord should be infiltrated in the region of emasculation as well. A nose twitch may also help with this method of restraint.

3) If a general anesthetic agent is not available the horse may be cast with the help of sedatives and ropes. Once the sedative has taken affect, the horse is cast by the use of two side lines. Once the horse is down, hobbles are placed on the front legs and the legs are then tied up to the chest. The hind legs are held in position by the two sidelines. Someone also needs to hold the head down when using this method.

4) This is the last and least recommended option and if used should be on young, manageable horses. When sedatives and general anesthetics agents are not available the horse is cast with two side lines. Once down on the ground hobbles are placed on the front legs and the front legs are tied up to the chest. The hind legs are held in position by the sidelines. Thismethod needs at least five people, four to hold the horse and one to perform the surgery.

All ropes used for casting and hobbles should be cotton or animal fiber (llama fiber, wool, etc.). Nylon ropes can result in rope burns that can be as harmful as any laceration.

CASTRATION STEPS

1) Castration is performed through incisions over each testicle. The incisions are located one centimeter off the center of the scrotum. The lower testicle is held between the thumb and fingers of one hand while an incision, the length of the testicle, is made with the other. The incision should go through the skin and subcutaneous tissues so that the testicle will pop out of the scrotum.

2) Any tissues, such as fat and subcutaneous tissues, should be stripped away from the testicle and the spermatic cord. The spermatic cord includes the vein, artery, cremaster muscle, and the ductus deferens.

3) Once the testicle and approx. six inches of the spermatic cord are exposed, the spermatic cord may be severed. This is best accomplished with an emasculator. An emasculator will cut the cord from the testicle while crushing the tissues on the cut end of the cord. Crushing the vein and artery reduces the likelihood of excessive bleeding. The emasculator should remain in place for one to two minutes. Care must be taken to hold the emasculator properly or the tissues on the wrong side of the cut will be crushed.

In the event that an emasculator is not available the cord may be crushed with a large hemostat or even between blocks of wood. Some individuals cauterize the end of the cord with a hot iron if an emasculator is not available.

4) Repeat the procedure on the other testicle.

5) Once the testicles have been removed the skin incisions are enlarged, by the use of fingers or scissors, so that a 10 centimeter opening is made.

6) With the proper dose of ketamine or thiamylal sodium the horse will be anesthetized for 10 to 20 minutes. Then the horse will begin to recover. The horse should be kept down until it has recovered enough to stand without struggling and falling. To keep the horse down, one may need to cover the horse's head and eyes with a towel and hold the horse's neck and head down.

7) Antibiotics are recommended for five to seven days. Care should be taken to prevent screw worm infestations. The horse should be exercised twice a day until healing is complete.

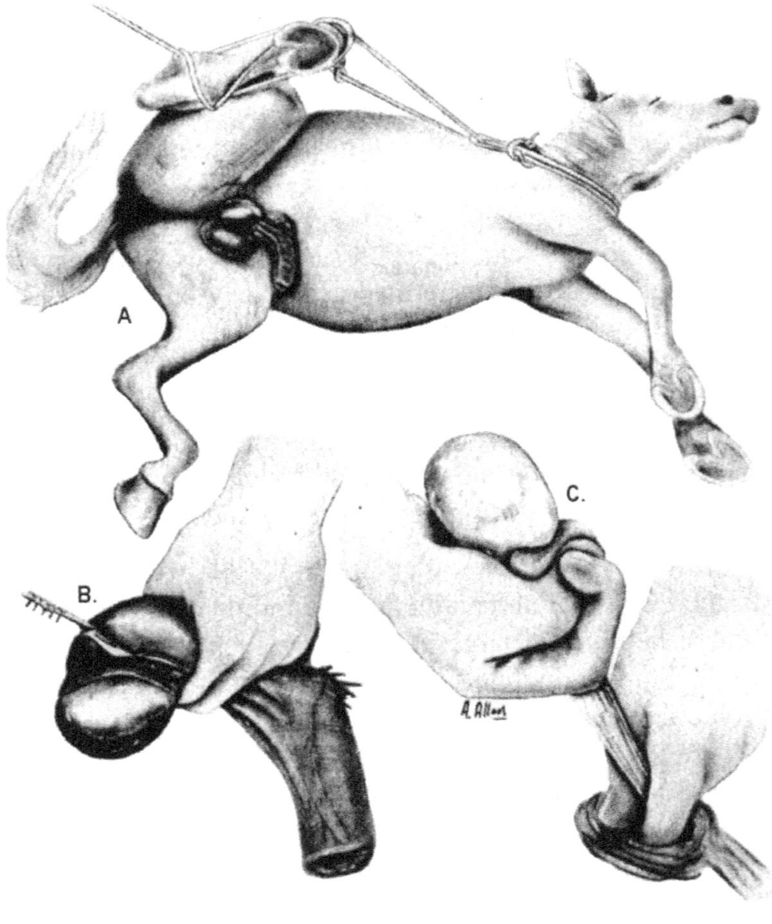

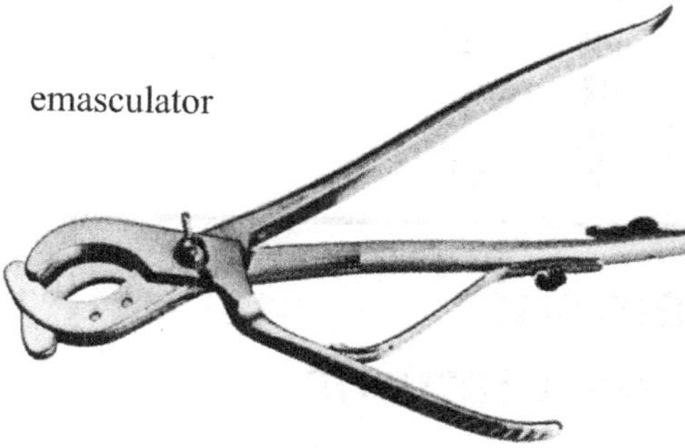
emasculator

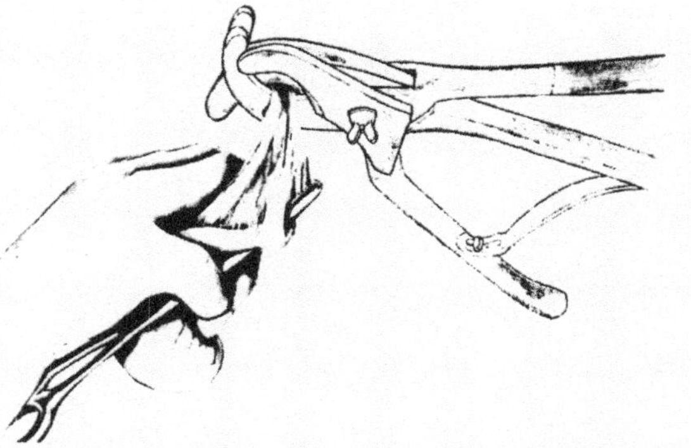

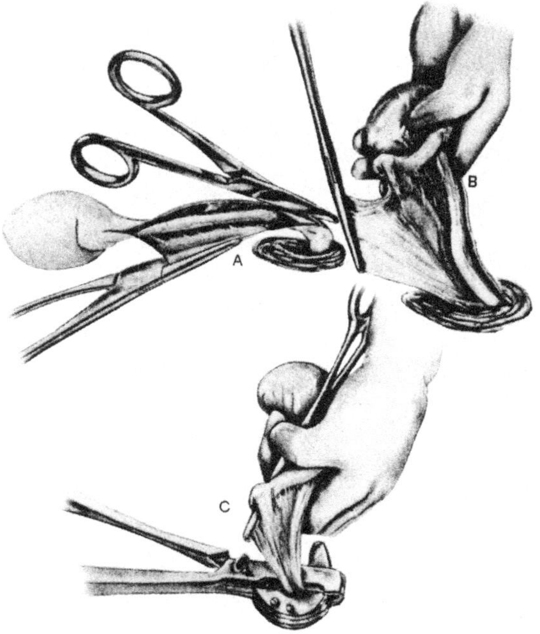
A
B
C

Section 15

Management Recommendations

Vaccinations

Vaccination	Frequency
Adult Horse Recommendation:	
Encephalitis (EEE, WEE, VEE)	Prior to fly season
Tetanus	12 months or if wounded
Influenza	6 months
Rhinopneumonitis	6 months
Vaccination for Pregnant Mares:	
Rhinopneumonitis	5th, 7th, 9th month of preg.
Encephalitis, Tetanus, Flu	One month prior to foaling
Foals:	
Encephalities, Tetanus, Flu & Rhino	At 4 & 6 months of age then to adult schedule
Suggested if disease is in the area & vaccine available:	
Rabies	12 months
Anthrax	12 months
Equine Viral Arteritis	12 months
Strangles	12 months

Deworming

Adult Horse Recommendation:
Deworm horses at least four times per year.

- Ivermectin (Equvalan or Zirmectrin paste or liquid) given orally works well after hard frosts and to kill larvae in the skin during the hot humid months. This dewormer may have to be given every month during the fly season to control fly-caused skin problems.
- Refer to Section 8 on parasites for specific product use, and on medications see section 18 for dosages.

Pregnant Mare Recommendation:

- Deworm on adult horse schedule.
- Deworm with Ivermectin one month prior to foaling.
- Deworm the day of foaling with Ivermectin.

Foal Recommendation:

- Deworm at two months of age.
- Deworm every two months until 18 months of age then use the adult schedule.

Hoof Care

- Hooves need to be trimmed every 6-8 weeks to prevent cracking.
- Shoes should be changed or reset every 8 weeks.

Dental Care

- Teeth should be examined every year, or when the horse has problems eating, chewing, or appears thin.
- Float teeth as needed.
- Caps may need to be removed between 2 and 3 1/2 years.
- Remove wolf teeth before using a bit in the mouth.

Disposal of Carcass

If an animal dies, do not feed its intestinal tract and contents to dogs. If the animal is suspected of having an infectious dis- ease, the carcass should be burned and the remains buried if local custom will allow—not even the hide should be saved from these animals as it can also spread the disease. Burial is another alternative to prevent the spread of infection. Burial should be at least 4 feet below the surrounding surface at a sitewhere the water level is at least 10 feet below the surface. Burial should also be over 500 feet from a stream or body of water.

Decontamination of the area with cleaning of manure and any other organic debris is very important. The area should then be cleaned before disinfection.

Common Disinfectants:

Formaldehyde	toxic fumes good for killing Anthrax on wood surfaces
Iodine	need prolonged contact does not kill Anthrax spores
Chlorine	not effective when organic material (blood, manure, etc.) is present can be caustic or corrosive

Section 16

Medications

Routes of Administration (methods of giving medications)

1. **Orally** - given via the mouth
   - **Do not** give large pills or boluses orally to a horse.
   - Capsules, pills, or boluses can be crushed, mixed with syrup and given through a deworming syringe orally.
   - Powders can also be given with syrup as above.
   - Horses will eat some oral medications in their feed if a sweet syrup or molasses is given to hide the medication's taste. Applesauce will also work to hide the taste.
   - Liquids should be given through a stomach tube passed through the nostril and into the esophagus. Only experienced people should attempt to do this.
   - Liquids can be mixed in the drinking water if they are water soluble and do not taste bad.
   - Liquids can also be poured (drenched) down the mouth of the horse, although some of the liquid may be lost.
2. **Intravenous (IV)** - given directly into the vein
   - This must be given intravenously, usually in the jugular vein of the neck.
   - Care must be taken to avoid the carotid artery which lies below the jugular (if blood pulses out of the needle, you are in the artery).
   - Be sure to inject directly into the vein as some medications can cause severe irritation if injected outside the vein.
3. **Intramuscular (IM)** - given in the muscle
   - Give in the middle of the neck or the thigh muscle
   - Never inject in the hip muscle of the horse, or near the shoulder.
   - Insert the needle first; if no blood appears, attach the syringe and pull back on the plunger. If no blood appears, give the injection.
   - Some medications cause soreness so you will need to use a new injection site each day.

4. **Subcutaneous (SQ) -** given by injection under the skin
   - Very few injections are given to the horse by this method.
   - Absorption rates may be slower from this method.
5. **Intra-articular -** given by injection into the joint
   - These injections should ONLY be given by those specially trained to do so.
6. **Topical Application -** rubbed onto the skin
   - Be sure the area is clean (free from dirt and debris) before topical medication is applied.
   - Some topical medications can blister the skin of sensitive individuals. Use caution!

Antibiotic - Antifungal

Drug	Dosage	Route	Frequency
Ampicillin	3 gm/ 500kg	IV-IM	12 hr
Cefazolin	2 mg/kg	IV-IM	12-24 hr
Erythromycin	4-8 mg/kg	IM	12-24 hr
Gentamicin	2.2 mg/kg	IV-IM	12 hr
Griseofulvin	5-10 mg/kg	Oral	24 hr
Iodides (Sodium)	125ml of sodium solution	IV	daily x 3
Kanamycin	4 mg/kg	IM	8 hr
Metranidazole	15-25mg/kg	Oral	12 hr
Naxcell	1 mg/kg	IM/IV	12 hr
Neomycin	5 mg/kg	Oral	12 hr
Oxytetracycline	5 mg/kg	IV	12-24 hr
Penicillin G, procaine	20-30,000 IU/kg	IM	12-24 hr
Penicillin, Benzathine	20-40,000 IU/kg	IM	48-72 hr
Rifampin	5 mg/kg	Oral	8 hr
Spectinomycin	20 mg/kg	IM	8-12 hr
Sulfathizole	66 mg/kg	Oral	8 hr
Sulfadiazine	50 mg/kg	Oral	12 hr
Sulfadimethoxine	27.5 mg/kg	Oral	12 hr
SMZ-TMP (Trimethoprim-sulfa)	24 mg/kg	Oral	12 hr

Antiprotozoals

Drug	Dose	Route	Frequency
Imidocarb for *B caballi*	2 mg/kg	IM	daily x 2 days
Imidocarb for *B equi*	5 mg/kg	IM	72 hours x 4 inj
Quinapyramine deimethyl sulfate or chloride	5 mg/kg	SQ	daily

Antifungal

Griseofulvin	10 mg/kg do not use in pregnant animals	orally for 7 days

Anesthetics

Acepromazine	0.04-0.1mg/kg	IV/IM	tranquilizer
Detomidine	0.02-0.04mg/kg	IV	tranquilizer
Thiamylal Sodium	1 gram for 500kg	IV	light general
Xylazine HCL	1.1mg/kg 2.2mg/kg	IV IM	tranquilizer

Anthelmintics (Anti-Parasitic)

Febendazole (Panacur)	5mg/kg orally 10 mg/kg orally for 5 days to kill larvae
Ivermectin (Eqvalan)	0.2 mg/kg orally
Mebendazole (Telmin)	8.8 mg/kg orally
Oxibendazole (Anthelcide)	10 mg/kg orally
Pyrantel Pamoate (Strongid)	12.5 mg/kg orally
Thiabendazole (TBZ)	50-100 mg/kg orally

Nonsteroidal Anti-inflammatory Agents

Asprin	15-100 mg/kg daily orally
Flunixin Meglumine (Banamine)	1.1 mg/kg orally daily 1.1 mg/kg IV of liquid
Meclofenamic Acid (Arquel)	2.2 mg/kg orally daily
Phenylbutazone (Bute)	4.4-8.8 mg/kg daily orally 1-2 grams IV daily

Gastrointestinal Agents

Laxatives

Docusate (DSS)	10-20 mg/kg in 2 liters of water given by stomach tube
Magnesium Salts	0.2 gm/kg diluted in 4 liters of water given by stomach tube if animal is well hydrated
Mineral Oil	2-4 liters via stomach tube
Psyllium	0.5 kg in 6-8 liters of water via stomach tube 125 gm mixed with grain daily

Antidiarrheals

Bismuth Subsalicylate	1 oz per 8 kg every 8 hours orally
Kaolin/Pectin	1 oz per 8 kg every 6-8 hours orally

Hormones

Altrenogest (Regumate)

1 ml per 50 kg once daily to keep out of heat

Use daily for 15 days to synchronize heat cycles.

Use daily to maintain pregnancy in mares that have low progesterone levels up to day 150 of pregnancy.

Oxytocin — 40 units IM at 60 minute intervals for retained placentas

20 units 3-4 times daily for 2-3 days for infected uterus (metritis)

Prostaglandin F2 alfa (Lutalyse)

5mg IM to induce heat cycles if greater than 5 days since last heat

Steroids

Dexamethasone — 2.5-5 mg IV or IM

Prednisone — 0.24-1 mg/kg orally daily or every other day

Weights and Measures:

.3937 inch = 1 centimeter

1 inch = 2.54 centimeter

12 inches = 1 foot = .3048 meter
3 feet = 1 yard = .944 meter
39.37 inches = 1 meter
5280 feet = 1 mile = 1.609 kilometers

1 acre = 43,560 square feet = .405 hectares
1 acre = 4,840 square yards = 4.05 hectares
2.74 acres = 1 hectare

to change Centigrade to Fahrenheit, multiply by 9/5 & add 32
to change Fahrenheit to Centigrade, subtract 32 & multiply by 5/9

1 milligram = .015 grains
1 ounce = 437.5 grains = 28.350 grams
1 pound = 453.592 grams
1,000 grams = 1 kilogram = 2.205 pounds
2,000 pounds = 1 net or short ton
2,204 pounds = 1 metric ton

16 ounces = 1 pint = .473 liters
2 pints = 1 quart (US) = .946 liter
4 quarts = 1 gallon (US) = 3.785 liters

1000 milliliters (ml) = 1 liter
1 milliliter = 1 cubic centimeter (cc)
1 teaspoon = 5 milliliters = 1/3 tablespoon
1 tablespoon = 15 milliliters = ½ ounce
8 ounces = 1 cup

Section 17

Glossary of terms

Abscess: a pus-filled cavity resulting from inflammation and usually caused by bacterial infection.
Accumulation: the process of gathering together and increasing in amount over a period of time.
Acute: extremely serious, severe, or painful.
Anemia: a blood condition in which there are too few red blood cells or the red blood cells are deficient in hemoglobin, resulting in poor health.
Auscultation: the act of listening to the sounds made by a patient's internal organs, especially the heart, lungs, and abdominal organs, usually with a stethoscope, in order to make a diagnosis.
Bight: a loop or slack curve in a rope.
Bolus: a soft rounded ball, especially of chewed food. A very large pill.
Booster: a repeat dose of a vaccine given some time after the initial course to maintain the level of immunity provided by the previous dose. A supplementary dose of vaccine.
Bots: a parasitic disease of horses, sheep, and cattle, caused by infection with botfly larvae. The larval stage of the botfly may be called bots.
Britchen/Breeching: the rear part of a harness that goes behind the buttocks/tail of the horse.
Castrate/Castration: to remove the testicles of a man or male animal, making reproduction impossible. Animals are sometimes castrated to make them more docile and to prevent disease.
Causative agent: something that makes something else happen; the reason it happened.
Caustic: a substance that can corrode or burn away other substances by chemical action.
Cavity: a hole or hollow space in something.
Compost: a mixture of decayed plants and other organic matter used by gardeners for enriching soil.
Condition: to make people or animals act or react in a particular way by gradually getting them used to a specific pattern of events; to give a treatment to improve general health, soundness, readiness for use, appearance, or performance.
Conformation: the shape, outline, or form of something, especially an animal, determined by the way in which its parts are arranged.
Contaminate: to make dirty or impure, make unfit for use; to make something such as soil unfit for use or exploitation as a result of contact with polluting or harmful substances.
Cribbing/crib-biting: a behavioral pattern that develops in horses kept in stables, marked by chewing of the stalls and swallowing air. They do this because they are bored.
Crupper: a strap that passes under the tail of a horse and is attached to a saddle or harness to prevent it from sliding forward.

Culture: the growing of biological material, especially plants, microorganisms, or animal tissue, in a nutrient substance culture medium in specially controlled conditions for scientific, medical, or commercial purposes.
Cutaneous: relating to the skin.
Debride: removal of dead and decaying tissue surrounding a wound.
Debris: fragments of something that has been destroyed or broken into pieces.
Decontaminate/Decontamination: to remove unwanted chemical, radioactive, or biological impurities or toxins from a person, object, or place. Cleanse, purify, clean up, disinfect…
Degeneration: the process of becoming physically, morally, or mentally worse.
Diagnostic: a test, procedure, or instrument used to identify the nature or cause of an illness, disorder, or problem.
Differentiate: to see or show the differences between two or more things.
Dilute: to make something thinner or weaker by adding water or another liquid.
Discharge: to emit, give off, or dispose of a gas or liquid, or be emitted or disposed of, i.e. nasal discharge.
Disinfect/Disinfection: to clean something so as to destroy disease-carrying microorganisms and prevent infection.
Dystocia: difficult birth or delivery.
Dorsal: relating to or situated on the back of the body.
Ectoparasites: Parasites that live on the outside of the animal and attack the skin and body openings.
Equine: a horse or other member of the horse family; includes horses, mules, zebras, and donkeys.
Esophagus/esophageal: the passage down which food moves between the throat and the stomach.
Estrus: a regular period of sexual excitement in many female mammals, during which the animal seeks to mate.
Evaluate: to consider or examine something in order to judge its value, quality, importance, extent, or condition.
Everting: to turn an organ or other body part outward or inside out.
Excessive: beyond what is considered acceptable, proper, usual, or necessary. Extreme.
Exostosis: a benign (not cancer) bony growth on the surface of a bone or a tooth root, caused by inflammation or repeated trauma.
Flexion: the bending of a limb or joint.
Flush: to clean or clear something by liberally pouring water or another liquid into, on, or through it.
Foreign body: something that is not natural to the horse; something that does not belong.
Fusion: the merger or a blending of two or more things such as materials or ideas.

Gestation: the process of carrying offspring in the womb during pregnancy.
Hames: either of a pair of metal or wooden bars curved to fit over the neck of a draft animal and to which the traces are attached.
Heat/Estrus: a time during a female mammal's reproductive cycle when she is fertile and ready to mate.
Hemorrhage: the loss of blood from a ruptured blood vessel, either internally or externally.
Hitch: a device used to connect two things, e.g. a ball on a vehicle for connecting a trailer.
Hobbles/Hopples: something, e.g. a loop of rope or a strap, used to tie the legs of a horse.
Host: a human, animal, plant, or other organism in or on which another organism, especially a parasite, lives.
Incubate: to keep cells or microorganisms at a controlled temperature in or on a medium so that they multiply.
Infest (infestation): to overrun a place or site in large numbers and become threatening, harmful, or unpleasant.
Infiltrate: to pass through a substance by filtration, or make a liquid or gas pass through a substance by filtration (i.e. inject).
Inflammation: swelling, redness, heat, and pain produced in an area of the body as a reaction to injury or infection.
Ingest/ingestion: to take food, liquid, or some other substance into the body by swallowing it.
Incision: a cut or the act of cutting, especially when performed by a surgeon.
Intermittent: occurring at irregular intervals; sometimes it is there, sometimes it is not there.
Intra-articular: into the joint.
Lame/lameness: walking unevenly because of a leg injury or motion impairment.
Laceration: a cut or gash in the skin where the wound is deep with irregular edges.
Lateral: relating to, located at, or affecting the side; a part, appendage, movement, or object at the side of something.
Lesions: a wound, especially an area of skin that is broken or infected.
Mange: a parasitic skin disease of animals and sometimes humans that is caused by mites and results in hair loss, scabs, and itching.
Medial: situated in or toward the middle.
Membranes: (mucous membranes) underside of the lip, gums, or the lining of the nose or eye.
Microfilaria: the early larval stage of a parasitic nematode worm filaria, a cause of heartworm in dogs and elephantiasis in humans.
Motility: capable of or demonstrating movement by independent means.

Nutrition: food, foods, or the minerals, vitamins, and other nourishing substances that they contain; the process of absorbing nutrients from food and processing them in the body in order to keep healthy or to grow.
Obstruction: something that causes or forms a blockage or hindrance; something that stops the normal passage or flow of something else.
Palpate/palpation: to examine a part of the body by feeling with the hands and fingers.
Parameters: facts or circumstances that restrict how something is done or what can be done. Distinguishing features or notable characteristics.
Parasite Load: number of parasites in an area. High parasite load means a lot of parasites are in that area (or animal).
Parturition: the act of giving birth to offspring.
Penetrate/penetration: to enter or pass through something, e.g. by piercing it or forcing a way in.
Predispose: to make somebody liable or inclined to do something such as catch an illness or behave in a particular way.
Procedure: an established or correct way of doing something.
Prognosis: a medical opinion as to the likely course and outcome of a disease.
Protectant: a substance that prevents something from being damaged, e.g. a coating used to keep metal from rusting.
Pus: the yellowish or greenish fluid that forms at sites of infection, consisting of dead white blood cells, dead tissue, bacteria, and blood serum.
Quarantine: to isolate (keep apart from others) a person or animal that may have been exposed to a contagious or infectious disease in order to prevent the possible spread of that disease.
Radiograph/x-ray: an image produced on film or another sensitive surface by radiation such as X-rays or gamma rays passing through an object.
Recumbent: lying back or lying down.
Resistance: when an animal or parasite no longer responds to a medicine, it has usually built up a resistance to that medicine and another medicine must be used.
Respiratory: relating to or used in breathing or the system in the body that takes in and distributes oxygen.
Rostral: toward the nose.
Rump: the fleshy hindquarters of a four-legged mammal, not including its legs (bottom/buttocks).
Scaliness: covered in scales or flakes.
Silage: animal fodder that is made by storing green plant material in a silo where it is preserved by partial fermentation.
Sign/Symptom: an indication of a disease or other disorder, especially one experienced by the patient, e.g. pain, dizziness, or itching.
Singletree/Whiffletree: a horizontal crossbar used to attach the harness traces of a draft animal, that is then attached to a vehicle or device (to be pulled by the horse).

Slough: a layer of dead skin that separates from healthy skin after an infection or inflammation.
Solar horn: sole of the horse's foot.
Soothant: something used to make pain or discomfort less severe.
Strongyle: a parasitic nematode worm related to the hookworm that infests the intestinal tract of mammals. Superfamily: Strongyloidea
Superficial: relating to, affecting, or located on or near the surface of something; not deep.
Supportive care: make sure they are comfortable, well-fed, and relatively painless.
Susceptible: easily influenced or affected by something.
Suture: a piece of material used to close a wound or connect tissues, e.g. catgut, thread, or wire. (Verb) To close a wound by joining the edges
Syndrome: a group of signs that together are characteristic or indicative of a specific disease or other disorder.
Synovial Fluid: a clear viscous fluid that lubricates the linings of joints and the sheaths of tendons.
Tack: saddles, bridles, and other parts of a horse's harness.
Terrets: either of two metal rings attached to the driving harness of a horse, through which the reins are passed to prevent them from slipping around the horse's flanks.
Topical: describes drugs or medications that are applied directly to the surface of the part of the body being treated.
Toxic: involves something poisonous; deadly; causing serious harm or death.
Toxin: a poison produced by a living organism; a substance that accumulates in the body and causes it harm.
Udder: a bag-shaped structure containing two or more milk-secreting glands, each with its own teat, found in mammals such as cows, sheep, and goats.
Unpigmented: having no natural color.
Ventral: located on or affecting the lower surface of an animal's body, or the front of the human body, relating to or situated in, on, or near the abdomen.
Via: by way of or through. By means of, using, by…
Viral: relating to, typical of, or caused by a virus.
Welts: raised ridges or bumps on the skin,
Withers: the bony ridge between the shoulder bones of a horse, sheep, ox, or similar four-legged animal, forming the highest part of its back.

Most definitions taken from the Encarta Dictionary of North American English

Section 18

Acknowledgements

Special Acknowledgments:

Dr. Leroy Dorminy - CVM Founder for resource information, suggestions, patience, and prayer.

Dr. Kelly Ward - Artist, drawings in this publications

Other Artists - Kelly C. Ward, DVM
Jessica Evans
Shannon Smith
Betsy and Paul Hutchins (The Definitive Donkey)
George Hatley (Horse Camping)

CVM Fieldworkers - Observations of problems in the field in developing countries

Dr. J.C. Shaul - Editing and suggestions

Dr. Jack Amen - Author, Surgery Chapter, editing and suggestions

Dr. Monty Mathis - Editing and suggestions

My wife, Susan - For patience, prayer, and loving support

Donors:

The Larry Barnes Foundation
Dr. Cameron Shaul
Collierville Animal Clinic

Bibliography

Back, Joe. (1994). Horses, Hitches and Rocky Trails. Johnson Books.

Equine Research Inc. (1994). The Illustrated Veterinary Encyclopedia for Horsemen. Equine Research.

Hatley, George B. and Portnoy, Lewis. (1992). Horse camping. The Appaloosa Museum.

Hadfield, Marabel. (1988). The manual of horsemanship. British Horse Society.

Kaplan RM. *"These ain't your father's parasites: An evidence-based medical approach to equine parasite control" The Practitioner* October 2008.

Leahy, John R. and Barrow, Pat (1953). Restraint of Animals. Cornell Campus Store.

Mansmann, R.A. (1982). Equine Medicine & Surgery. American Veterinary Publications.

Smith, Carin A. (1991). Easy health care for your horse: a veterinarian's approach to prevention management, and care. Prentice Hall Press Equestrian Books.

Yakstis, Jerome J. and Johnstone, Colin. (1981). Parasites of horses. Merck.

Webber, Toni. (1990). Feet and Shoes. Kenilworth Press.

Other Sources of Information

Equine Reproduction, 1991
Lameness in Horses, Stashack, Turner
Large Animal Internal Medicine, 1990, Smith
The Illustrated Veterinary Encyclopedia, 1975, Equine Research
The Merck Veterinary Manual, 1986, Merck & Co

Section 19

Deworming Addendum

Deworming strategies for the next generation:
what you need to know to protect your horse

Resistance is real. Equine parasite resistance has been reported in 21 countries, including the United States, and no new deworming drugs are currently in development. Keeping the effectiveness of the drugs we have for as long as possibleis therefore very important. Some European countries are so concerned about parasite resistance, in fact, that dewormers are once again only available through a veterinarian or by prescription.

Preventing transmission is key. Delaying parasite resistance and preventing clinical disease due to parasites are our primary goals. Today's deworming program should focus on preventing pasture contamination and transmission of infective larvae in large numbers.

Identify high shedders. Conventional deworming strategies over-treat many horses, but under-treat high shedders. High-shedding horses have less natural immunity to internal parasites than others. These horses shed many more eggs thanthose with higher immunity (known as low shedders) because between deworming treatments, their body allows more adult worms to survive and produce eggs. While high shedders only account for 20% of the horse population, they produce 80% of the parasite eggs on the pasture. High shedders need to be dewormed more often than their low shedding herdmates to prevent contamination of the pasture with large numbers of parasite eggs. Low shedding horses may only require deworming twice a year, depending on their management situation. A simple test called a **Fecal Egg Count** will identify high shedding vs. low shedding horses. This test is best performed at least 3 months after the last deworming treatment. In our climate, spring is a convenient time to perform the test because most horses are dewormed less in the winter.

Check for resistance. Small strongyle resistance to the benzimidazole (Panacur, Anthelcide) and pyrantel pamoate (Strongid) dewormer classes and ascarid resistance to macrocyclic lactone (Ivermectin, Quest) dewormer class have been reported in the United States. With this in mind, it is important to check for resistance periodically in your herd. A Fecal Egg Count Reduction Test is the preferred method for testing for resistance. This test is basically two Fecal Egg Counts; one done before deworming and another two weeks after deworming.

A few worms are OK. The goal of previous deworming strategies was to eliminate all parasites from the horse. In an effort to delay widespread resistance, however, today's deworming strategies should be focused on reducing transmission of parasites, keeping parasite burdens below harmful levels, and treating clinically affected horses. In this manner, a population of susceptible parasites is maintained to dilute the resistant genes in the population and thereby keeping our deworming medications effective as long as possible.

Young horses require special consideration. The primary parasite species of concern in young horses is the ascarid, or round worm. Most horses develop complete immunity to this parasite by the time they are 18 months old, but it causes significant damage in the meantime if left unaddressed. Unfortunately, ascarids have developed a high prevalence of ivermectin/moxidectin resistance on some breeding farms. Since resistance to the benzimidazoles has been seen in small strongyles, a species of concern in both young and adult horses, treatment with both classes of dewormers is sometimes necessary in young horses. For this reason, Fecal Egg Count Reduction Tests to check for resistance in foals and young horses are extremely important.

Deworming Recommendations

(Adjust to local calendar)

Adults

High Shedders:

April 1: Fecal Egg Count, treat with Ivermectin
June 1: Treat with Quest
Sept 1: Fecal Egg Count, treat with oxibendazole (Anthelcide), need Fecal Egg Count Reduction Test the first year to test for resistance
October or November, after a hard frost: treat with Ivermectin with Praziquantel

Low Shedders in High Density situation:

April 1: Fecal Egg Count, treat with Ivermectin
June 1: Treat with Quest
Sept 1: Fecal Egg Count, treat if egg count is high
October or November, after a hard frost: treat with Ivermectin with Praziquantel

Low Shedders in Low Density situation:

April 1: Fecal Egg Count, treat with Ivermectin
Sept 1: Fecal Egg Count, treat if egg count is high
October or November, after a hard frost: treat with Quest Plus

Foals

Fecal Egg Count Reduction Tests (FECRT) should be performed with all deworming medications used in foals to detect resistance common in ascarids and small strongyles. If no resistance is detected, adhere to the following plan. If resistance is detected, contact your veterinarian for advice. Once resistance pattern on farm is established, FECRT may only need to be done every few years.

2 months of age: Pyrantel Pamoate (Strongid) treatment (too young for FECRT)
4 months of age: FECRT in conjunction with Ivermectin treatment.
6 months of age: FECRT in conjunction with Pyrantel Pamoate (Strongid) treatment
8 months of age: FECRT in conjunction with Oxibendazole (Anthelcide) treatment
10 months of age: Ivermectin treatment
12 months of age: Start on Adult program, but monitor Fecal Egg Counts to detect Ascarid infection due to late-developing immunity

Submitted by Dr. David Wright and Dr. Heather D. Damico, Buffalo Equine and Large Animal Clinic, Buffalo, MN. Written from information from Kaplan RM. *"These ain't your father's parasites: An evidence-based medical approach to equine parasite control"* *The Practitioner* October 2008.

Drug Name	Egg Reappearance Period	Effective against	Resistance reported to
Ivermectin	8 wks	-Adult and larval (not encysted) Lg/Sm Strongyle, Ascarids, pinworms -ectoparasites (mites, lice, ticks, bots, cattle grubs) Ineffective against tapes and flukes	Ascarids on some breeding farms
Moxidectin	12 wks	-Adult and larval Lg/Sm Strongyles, late L3/ and L4 encysted small strongyles, Ascarids, pinworms	Ascarids on some breeding farms
Benzimidazoles Fenbendazole Oxibendazole	4 wks	-Adult small strongyles -L3/L4 encysted small strongyles when given at double dose for 5 days -Adult Ascarids and pinworms at double dose **Oxibendazole more effective in one dose than fenbendazole**	Large Strongyles Small Strongyles
Pyrantel	4 wks	-2.65 mg/kg daily dose (Strongid C) – ingested strongyle and ascarid larvae (not migrating) -6.6 mg/kg dose – adult lg/sm strongyles, adult ascarids -13.2 mg/kg dose - tapes	Small Strongyles Ascarids
Praziquantel	?	Tapes	none

Christian Veterinary Mission (Publisher of this book)

Our vision is to see

Christ's love expressed through veterinary medicine.

Our mission is to

challenge, empower and facilitate veterinarians to serve through their profession, living out their Christian faith.

CVM also provides education and encouragement for those who desire to minister through service, prayer, relationship building, and modeling Christ's love.

About CVM

Christian Veterinary Mission (CVM) is a registered non-profit Christian Service Organization 501(c)(3) based in Seattle, Washington, U.S.A.

CVM was founded in 1976 by Dr. Leroy Dorminy who came to realize the impact that veterinarians could have by integrating their faith with their practice, both locally and around the world. In 2008, CVM had nearly 30 veterinary professionals serving full-time internationally and over 200 veterinary professionals and student volunteers serve on short-term cross-cultural mission trips annually. CVM sponsors fellowship & prayer breakfasts at over 20 U.S. veterinary meetings each year and reaches out to veterinary students through Christian Veterinary Fellow-ship (CVF) groups in every veterinary school in the U.S. by encouraging them in spiritual growth and professional development.

There are over 3,500 veterinarians affiliated with CVM in the U.S. CVM also partners with organizations and networks in other countries that are focused on empowering Christian veterinarians. CVM has a volunteer advisory board of veterinarians who guide its vision, mission, and programming.

CVM books and the free International Animal Health Newsletter were written with small farmers, veterinarians, and agricultural development workers in mind. Our desire is that they would help individuals and groups develop an appropriate livestock program to meet community needs. CVM's Endowment Fund was started in the early years of the organization's life. The fund provides for meaningful programs that could not be funded by the regular budgeting process.

Made in the USA
Monee, IL
04 November 2026

67383943R00138